Technology Horizons

A Vision for Air Force Science and Technology 2010–30

Key science and technology focus areas for the US Air Force over the next two decades that will provide technologically achievable capabilities enabling the Air Force to gain the greatest US joint force effectiveness in 2030 and beyond

OFFICE OF THE US AIR FORCE CHIEF SCIENTIST

Originally released 15 May 2010 by the
United States Air Force Chief Scientist (AF/ST) as
*Report on Technology Horizons: A Vision for Air Force
Science & Technology during 2010–2030*
Volume 1
AF/ST-TR-10-01-PR

September 2011

Published by Books Express Publishing
Copyright © Books Express, 2012
ISBN 978-1-78039-784-9

Books Express publications are available from all good retail and online booksellers. For
publishing proposals and direct ordering please contact us at: info@books-express.com

Contents

Illustrations

Foreword

Technology Horizons is our vision for key Air Force science and technology investments over the next decade that will provide us with truly game-changing capabilities to meet our strategic and joint force responsibilities. The coming decades hold high promise for amazing new capabilities across the air, space, and cyber domains. Yet the Air Force and our nation will also be confronted with substantial strategic, technology, and budget challenges. Our greatest advances will come with a focused investment of resources in the most promising technologies. The vision in *Technology Horizons* provides the shared awareness of the challenges and opportunities that will enable us to achieve this focus.

Technology Horizons presents a clearly articulated and credible assessment of the strategic environment and enduring realities we face. It outlines a set of overarching themes that defines attributes our future Air Force systems will need to prevail. New technology-enabled capabilities are envisioned that meet key needs, including long-range strike, deterrence tools, cyber resilience, energy efficiency, and automation and enhanced human-machine interfaces, to help our most valuable asset—our Airmen—be even more effective than today. We believe the Air Force must boldly move forward to advance these technologies through the dedicated, creative, and focused efforts of our science, technology, engineering, and mathematics workforce. The future is ours to shape.

To implement this vision, we are concentrating a meaningful portion of our Air Force Research Laboratory effort on the identified key technologies. We will move forward in pursuing "grand challenges" that will help achieve militarily useful capabilities. We will work closely with our partners across the Department of Defense, government, industry, academia, and allied nations to leverage the best intellectual capital and facilities in pursuit of the most promising ideas. And we will sustain our focus on these science and technology efforts to maximize their likelihood of being transitioned into operational capabilities meeting Air Force needs. We firmly believe that maintaining our technical and operational superiority in this manner is both necessary and attainable.

Therefore, we encourage all Airmen—indeed all warriors and our other national and international partners—to read *Technology Horizons* and seriously contemplate the transformative opportunities that technology can enable in the coming decades. We call on you as Airmen to contribute your intellectual energy to developing new frameworks and novel concepts of operations to take maximum advantage of these coming technologies. These are challenging times, but we have no doubt that America's Airmen will overcome the challenges we face to provide the critical capability advances needed to ensure the United States Air Force remains the world's premier air force through 2030 and beyond.

Michael B. Donley
Secretary of the Air Force

Norton A. Schwartz
General, USAF
Chief of Staff

Preface

The proud heritage of the United States Air Force is closely intertwined with the advancement of science and technology (S&T), beginning with the invention of mechanical flight. In 1945, two years before the Air Force became a separate service, the world-renowned aerodynamicist Theodore von Kármán led our very first S&T vision, "Towards New Horizons," for Army Air Force general Henry "Hap" Arnold. Hap Arnold knew that "any air force which does not keep its doctrines ahead of its equipment, and its vision far into the future, can only delude the nation into a false sense of security." Von Kármán noted that "only a constant inquisitive attitude toward science and a ceaseless and swift adaptation to new developments can maintain the security of this nation." In "Towards New Horizons" he advocated new developments such as supersonic aircraft, air defense systems, and unpiloted vehicles, better known as the ballistic missiles that helped win the Cold War. Moreover, this groundbreaking work was the impetus for building the Air Force scientific laboratory system and test infrastructure without which today's capabilities could not have been realized. Compared to our adversaries at the start of World War II, the Air Force was transformed from a technologically inferior force to the world's most technologically advanced and formidable airpower within merely a decade.

Over a half century later, during the Gulf War, the Department of Defense (DOD) and the Air Force experienced the benefits of continued shaping of the future through applied S&T developments. A swift victory for the allied force with very limited civilian and coalition casualties was made possible by precision-guided munitions ("smart bombs") enabled by the satellite-based Global Positioning System (GPS), enhanced situational awareness enabled by the Joint Surveillance and Target Attack Radar System (JSTARS), superior command and control enabled by the Airborne Warning and Control System (AWACS), air superiority provided by F-117A Stealth aircraft, and air defense provided by Patriot missiles.

Today, in America's efforts to provide stability, security, and prosperity to weakened or failing states, the center of gravity has transitioned from military systems to civilian populations. Operational success is once again enabled by previous S&T investments and innovations.

These include remotely piloted aircraft (RPA), air/spaceborne multi-intelligence sensing (radar, infrared, signals intelligence), distributed command, control, and intelligence (e.g., distributed common ground system), micromunitions, directed energy, and human terrain modeling. Collectively, these have enabled conducting persistent surveillance (and thus have led to tactical patience), novel sensing of improvised explosive devices, tracking of deceptive adversaries who hide in plain sight, and protecting civilian populations, crucial in the battle for hearts and minds.

In spite of these breathtaking advances, change continues to accelerate. Last year alone, the Air Force added new career fields for cyber and RPA operators, driving new requirements for technology advances. What will determine success in future operational engagements remains unknown, but it will most certainly involve S&T. In the words of Air Force chief of staff Gen Norton A. Schwartz, "We cannot know what the future holds, so in order to realize my vision of a consistently powerful, capable Air Force, we will almost certainly need to pursue initiatives not yet fully imagined."

Today's multipolar, fiscally constrained, and technology-filled world demands vigorous and properly focused S&T investments to advance the capabilities that the Air Force will need to fulfill its mission in an uncertain future. Accordingly, the secretary of the Air Force and chief of staff of the Air Force requested that the Office of the Chief Scientist craft *Technology Horizons*. The former chief scientist of the US Air Force, Dr. Werner J. A. Dahm (October 2008 to September 2010), skillfully responded by orchestrating expert insights from a wide range of organizations, including the Air Force Research Laboratory, MAJCOMs, Air Staff, operational squadrons, sister services, DOD agencies, other federal agencies, federally funded research and development centers, national laboratories, industry, and academia.

Technology Horizons begins by characterizing the strategic context, noting today's contested, congested, and competitive multipolar environment and underscoring technology-derived challenges to Air Force capabilities in air, space, and cyber, including new threats in areas such as electronic warfare, directed energy, and GPS denial. Several enduring realities that shape what we can do as an Air Force are then expressed. These include not only budgetary constraints driven by growing human, sustainment, and energy costs but also opportunities to build upon legacy systems as well as strong partnerships with our sister services, DOD agencies, national laboratories, industry, and international partners. *Technology Horizons* next presents key overarching themes to help guide and focus S&T efforts. These include, for example, fundamental shifts toward agile, fractionated, and cross-domain systems operating in increasingly autonomous fashion in contested environments.

Recognizing the intimate link between capability and technology, *Technology Horizons* employed a unique methodology to chart the most productive technology pathways toward the highest value capabilities, including a focus on Air Force service core functions. The resulting potential capability and key technology areas allow both our operators and technologists to glimpse the art of the possible. Moreover, *Technology Horizons* provides an actionable plan to move forward toward that future. Technologists have in hand an expert assessment of those most promising areas in which to concentrate effort and a number of grand challenges that provide a context for beginning to integrate diverse technologies. Operators have a roadmap with which to begin considering future strategy, organization, and employment of these advanced capabilities and to influence the direction of technology development as it unfolds. For example, *Technology Horizons* identifies "disproportionately valuable" technology areas suited to the strategic, technological, and budget environments of 2010–30, such as flexible autonomy, human cognitive augmentation, cyber and spectrum resiliency, energy efficiency, and long-range strike.

The Air Force is at a pivotal time in its history. The confluence of strategic change, global technological advancement, and fiscal and natural resource constraints causes some to wonder how we will maintain our technological advantage. The importance of our technological advantage goes well beyond national security, as noted recently by Pres. Barack Obama: "Science is more essential for our prosperity, our security, our health, our environment, and our quality of life than it has ever been." The *Technology Horizons* vision enables the Air Force to vector its S&T investments over the coming decade to enable technologically achievable capabilities that can provide it with the greatest US joint force effectiveness by 2030. As Air Force secretary Michael B. Donley observes, "*Technology Horizons* will shape our future Air Force research and development priorities." Because of its clarity and focus, *Technology Horizons* will enable the Air Force to support the current fight and be responsive to Air Force service core functions while at the same time advancing break-through S&T for tomorrow's dominant war-fighting capabilities.

Our potential adversaries have not missed the powerful lessons of technological transformation and the advantages that accrue to an air force that embraces this mind-set. We must remain as committed as we were in 1945 to pursuing the most promising technological opportunities for our times, to employing the scientific and engineering savvy to bring them to reality, and to having the wisdom to transition them into the next generation of capabilities that will allow us to maintain our edge. In the words of General Schwartz, "Even as we focus on winning today's war, we should also keep a watchful eye on the evolving 21st century security environment. We must take steps today that will allow future generations to meet—and shape—the challenges of tomorrow. That will not be easy." While we face substantial strategic, operational, and economic challenges in the coming decade, *Technology Horizons* has laid out a perspicuous vision to guide the scientific and technological advancements necessary to sustain the operational superiority of our Air Force into the horizon and beyond.

Dr. Mark T. Maybury
Chief Scientist of the US Air Force
Pentagon

Acknowledgments

Technology Horizons was a major undertaking, and many people made important contributions to it. It is apparent that an effort of this scope, involving such a large number of people from so many different organizations for nearly a year, can only be undertaken about once every decade or so. With the process now completed and the final product delivered, it is appropriate to call out some of the key individuals who made among the greatest contributions to it.

Perhaps most noteworthy among these was Col Eric Silkowski, PhD. In addition to serving as my military assistant in AF/ST, he was a full participant in the fact-finding visits, briefings, and discussions involved in this effort. He also served as a member of all four of the *Technology Horizons* working groups and gave essential assistance with the writing and editing of the final report volumes. His technical background in engineering physics and his experience in numerous Air Force science and technology roles and relevant management positions were immeasurably important to this effort.

The MAJCOMs and product centers provided key inputs throughout this effort. Several individuals in those organizations made especially important contributions, particularly Gen Donald Hoffman and his staff in Air Force Materiel Command (AFMC), Gen Robert Kehler and his staff in Air Force Space Command (AFSPC), Gen Raymond Johns and his staff in Air Mobility Command (AMC), and Lt Gen Ted Bowlds and his staff at the Electronic Systems Center (ESC).

The Air Force Research Laboratory (AFRL) was an essential and highly active contributor. Maj Gen Curt Bedke and later Maj Gen Ellen Pawlikowski served as AFRL/CC during the course of the effort, and they made the staff in Headquarters AFRL and across the laboratory available to support the project in every way. Dr. Mike Kuliasha, AFRL's chief technologist at the time, was on all four working groups and made key contributions throughout the effort. Many individuals from AFRL's technical directorates served on the working groups; among these, Dr. Bill Baker, Dr. Alok Das, Mr. Jeff Hughes, Dr. Kamal Jabbour, Dr. Brian Kent, and Dr. Jim Riker must be given special mention. They and others across AFRL provided important background information and fact-finding materials. *Technology Horizons* would not have been possible without the strong contributions that AFRL made to it.

Colleagues on the Air Staff and the Air Force Secretariat provided important inputs and feedback at essential phases of the effort. Especially noteworthy contributions were made by Lt Gen David Deptula and his staff in AF/A2, Lt Gen Phil Breedlove and the AF/A3/5 staff, Lt Gen Chris Miller along with his staff in AF/A8, Dr. Jacqueline Henningsen and her staff in AF/A9, and Mr. Bruce Lemkin and his staff in SAF/IA.

Many others should also be noted for their contributions. Key among these are Dr. Tom Ehrhard of the Chief of Staff's Strategic Studies Group, Dr. Janet Fender in Air Combat Command, Dr. Don Erbschloe in AMC, Dr. Doug Beason in AFSPC, Dr. Mark Gallagher in AF/A9, Mr. Gary O'Connell and Ms. Betsy Witt at the National Air and Space Intelligence Center, Maj Gen Brad Heithold and his staff at the Air Force Intelligence, Surveillance, and Reconnaissance Agency, Dr. Ned Allen from Lockheed Martin, Dr. Richard Byrne from MITRE, and Dr. Richard Hallion. Lt Gen Bob Elder, USAF, retired, provided particularly important inputs on an early draft of the report.

The four working groups listed in appendix E were essential to *Technology Horizons*. Members of these groups not already noted above also deserve special mention. They include Dr. John Bay, Mr. Doug Bowers, Dr. Chris Colliver, Dr. Gregory Crawford, Dr. Roberta Ewart, Col Robert Fredell, PhD, Mr. Jon Goding, Mr. Jonathan Gordon, Dr. Thomas Hamilton, Dr. David Hardy, Mr. Dewey Houck, Mr. Richard Mesic, Dr. Lara Schmidt, Dr. Dwight Streit, Dr. Marc Zissman, and Dr. John Zolper.

Ms. Penny Ellis in AF/ST provided expert administrative support throughout the effort. Given the number of site visits, fact-finding trips, working group meetings, and briefings that were involved in producing *Technology Horizons*, this was no small task, but it was one that she performed with her characteristic skill.

Without the collective contributions made by these and many others, *Technology Horizons* could not have been possible and would not have reflected the broad range of inputs needed to develop a clear vision for the most essential Air Force science and technology focus areas.

Dr. Werner J. A. Dahm
Chief Scientist of the US Air Force (2008–10)
Pentagon

Executive Summary

From its inception, the Air Force has conducted a major effort roughly once every decade to articulate a vision for the science and technology (S&T) advancements that it should undertake to achieve over the following decade to enable the capabilities it will need to prevail. Six such S&T visions have been developed, beginning with *Toward New Horizons* in 1945 led by Theodore von Kármán for Gen Hap Arnold, through *New World Vistas* conducted in 1995.

Since completion of the latter, 15 years have passed without an updated Air Force S&T vision. *Technology Horizons* represents the next in this succession of major vision efforts conducted at the Headquarters Air Force (HAF) level. In view of the far-reaching strategic changes, rapid global technological advances, and growing resource constraints over the next decade, this is an overdue effort that can help guide S&T investments to maximize their impact for maintaining Air Force technological superiority over potential adversaries.

What Is *Technology Horizons*?

Technology Horizons is neither a prediction of the future nor a forecast of a set of likely future scenarios. It is a rational assessment of what is credibly achievable from a technical perspective to give the Air Force capabilities that are suited for the strategic, technology, and budget environments of 2010–30. It is visionary, but its view is informed by the strategic context in which these technology-derived capabilities will be used. It is an articulation of the "art of the possible" but is grounded in the knowledge that merely being possible is only a prerequisite to being practically useful. It considers the spectrum of technical possibilities but acknowledges that budget constraints will limit the set of these that can be pursued.

It recognizes that increasingly more of the science and technology that provides the basis for future Air Force capabilities is available worldwide to be translated into potential adversary capabilities. It thus has sought to envision not only US joint and allied opportunities for using technologies, but also ways that adversary capabilities could be derived from them using entirely different concepts of operations or on the basis of entirely different war-fighting constructs. It acknowledges that capabilities enabled by new technologies and associated operating concepts often introduce new vulnerabilities not envisioned in the original capability. It thus has also considered potential vulnerabilities and cross-domain interdependences that may be created by second- and third-order effects of these technology-derived capabilities.

Sources of Inputs

Technology Horizons received inputs from a wide range of organizations and sources. These included four working groups—one each in the air, space, and cyber domains and another that focused on cross-domain insights—that gave a broad range of subject matter expertise to this effort. Working group members were drawn from the Air Force S&T community, intelligence community, MAJCOMs, product centers, federally funded research and development centers (FFRDC), defense industry, and academia. Further inputs were obtained from site visits, briefings, and discussions with organizations across the Air Force, the Department of Defense, federal agencies, FFRDCs, national laboratories, and industry, including Air Staff and Air Force Secretariat offices. Additional use was made of perspectives in several hundred papers, reports, briefings, and other sources.

Elements of the Resulting S&T Vision

The vision from *Technology Horizons* to help guide Air Force S&T over the next decade and beyond consists of the following elements:

1. Strategic Context
2. Enduring Realities
3. Overarching Themes
4. Potential Capability Areas
5. Key Technology Areas
6. Grand Challenges
7. Summary of S&T Vision
8. Implementation Plan & Recommendations

Overarching Themes for Air Force S&T

The strategic context and enduring realities identified in *Technology Horizons* lead to a set of 12 overarching themes to vector S&T in directions that can maximize capability superiority. These shifts in research emphases should be applied judiciously to guide each research area.

1. From ... Platforms   To ... Capabilities
2. From ... Manned   To ... Remotely piloted
3. From ... Fixed   To ... Agile
4. From ... Control   To ... Autonomy
5. From ... Integrated   To ... Fractionated
6. From ... Preplanned   To ... Composable
7. From ... Single-domain   To ... Cross-domain
8. From ... Permissive   To ... Contested
9. From ... Sensor   To ... Information
10. From ... Operations   To ... Dissuasion/Deterrence
11. From ... Cyber defense   To ... Cyber resilience
12. From ... Long system life To ... Faster refresh

Potential Capability Areas, Key Technology Areas, and Grand Challenges for Air Force S&T

Based on the strategic environment, enduring realities, and overarching themes, the remaining elements of the vision from *Technology Horizons* are presented as follows to help guide the Air Force in making choices about the most essential S&T efforts that must be pursued to prepare for the environment of 2010–30. It identifies potential capability areas (PCA) and maps these PCAs across the Air Force service core functions to assess the range of impact they can have.

- It uses this set of PCAs to identify key technology areas (KTA) that are most essential for the Air Force to invest in over the next decade to obtain capabilities aligned with the strategic, technology, and budget environments.

- It additionally defines four "grand challenge" problems to advance KTAs and integrate them in system-level demonstrations of significant new capabilities.

- It presents an implementation plan that enables the elements of this vision to be put into practice for guiding Air Force S&T efforts to maximize resulting capabilities in 2030.

Major Findings

The single greatest theme to emerge from *Technology Horizons* is the need, opportunity, and potential to dramatically advance technologies that can allow the Air Force to gain the capability increases, manpower efficiencies, and cost reductions available through far greater use of autonomous systems in essentially all aspects of Air Force operations. Increased use of autonomy—not only in the number of systems and processes to which autonomous control and reasoning can be applied but especially in the degree of autonomy that is reflected in these—can provide the Air Force with potentially enormous increases in its capabilities and, if implemented correctly, can do so in ways that enable manpower efficiencies and cost reductions.

Achieving these gains will depend on development of entirely new methods for enabling "trust in autonomy" through verification and validation (V&V) of the near-infinite state systems that result from high levels of adaptibility and autonomy. In effect, the number of possible input states that such systems can be presented with is so large that not only is it impossible to test all of them directly, it is not even feasible to test more than an insignificantly small fraction of them. Development of such systems is thus inherently unverifiable by today's methods, and as a result their operation in all but comparatively trivial applications is uncertifiable.

It is possible to develop systems having high levels of autonomy, but it is the lack of suitable V&V methods that prevents all but relatively low levels of autonomy from being certified for use. Potential adversaries, however, may be willing to field systems with far higher levels of autonomy without any need for certifiable V&V and could gain significant capability advantages over the Air Force by doing so. Countering this asymmetric advantage will require as-yet-undeveloped methods for achieving certifiably reliable V&V. The Air Force, as one the greatest potential beneficiaries of more highly adaptive and autonomous systems, must be a leader in the development of the underlying S&T principles for V&V.

A second key finding to emerge from *Technology Horizons* is that natural human capacities are becoming increasingly mismatched to the enormous data volumes, processing capabilities, and decision speeds that technologies either offer or demand. Although humans today remain more capable than machines for many tasks, by 2030 machine capabilities will have increased to the point that humans will have become the weakest component in a wide array of systems and processes. Humans and machines will need to become far more closely coupled through improved human-machine interfaces and by direct augmentation of human performance.

Focused research efforts over the next decade will permit significant practical instantiations of augmented human performance. These may come from increased use of autonomous systems as noted above, from improved man-machine interfaces to couple humans more closely and more intuitively with automated systems, or from direct augmentation of humans themselves. The latter includes drugs or implants to improve memory, alertness, cognition, or visual/aural acuity, as well as screening of individuals for speciality codes based on brainwave patterns or genetic correlators, or even genetic modification itself. While some such methods may appear inherently distasteful, potential adversaries may be entirely willing to make use of them.

Developing ways of using science and technology to augment human performance will become increasingly essential for gaining the benefits that many technologies can bring. Significant advances and early implementations are possible over the next decade. Such augmentation is a further means for increasing human efficiencies, allowing reduced manpower needs for the same capabilities or increased capabilities with given manpower.

A further key theme is the need to focus a greater fraction of S&T investments on research to support increased freedom of operations in contested or denied environments. Three main research areas are of particular importance: (1) cyber resilience, (2) precision navigation and timing in Global Positioning System (GPS)-denied environments, and (3) electromagnetic spectrum warfare. Additionally, the study identifies further key priority areas where S&T investment will be needed over the next decade to enable essential capabilities, including processing-enabled intelligent sensors, directed energy for tactical strike/defense, persistent space situational awareness, rapidly composable small satellites, and next-generation high-efficiency gas turbine engines.

Recommendations

Technology Horizons makes five major recommendations for guiding Air Force S&T efforts to meet the strategy, technology, and budget challenges over the next decade and beyond:

Recommendation #1: Communicate Results from *Technology Horizons*.

- Communicate the rationale, objectives, process, and key elements from *Technology Horizons* via briefings offered to all HAF offices, MAJCOMs, product centers, and the Air Force Research Laboratory (AFRL).

- Build broad awareness, understanding, and support for the Air Force S&T vision from *Technology Horizons*.

- Disseminate *Technology Horizons* across all relevant organizations beyond those that provided inputs to this effort.

Recommendation #2: Assess Alignment of the S&T Portfolio with *Technology Horizons*.

- Assess alignment of AFRL's current S&T portfolio with the broad research directions and technology focus areas outlined in *Technology Horizons*.

- Identify the target fraction of the total Air Force S&T portfolio to be aligned with the research directions and technology focus areas identified in *Technology Horizons*.

Recommendation #3: Adjust the S&T Portfolio Balance As Needed.

- Identify research efforts in the current S&T portfolio that must be redirected or realigned with the research directions and technology focus areas in *Technology Horizons*.

- Determine which of these efforts should be realigned, redirected, or terminated to accommodate new research efforts that achieve the needed direction and emphasis.

- Define new research efforts that will be started to allow broad research directions and technology areas identified in *Technology Horizons* to be effectively achieved.

- Implement changes in the AFRL S&T portfolio to initiate new research efforts identified above and to realign, redirect, and terminate existing efforts identified above.

Recommendation #4: Initiate Focused Research on "Grand Challenge" Problems.

- Evaluate, define, and focus a set of grand-challenge problems of sufficient scale and scope to drive major technology development efforts in key areas identified here.

- Structure each grand challenge to drive advances in research directions identified in *Technology Horizons* as being essential for meeting Air Force needs in 2030.

- Define specific demonstration goals for each challenge that require sets of individual technology areas to be integrated and demonstrated at the whole-system level.

- Initiate sustained research efforts in the AFRL S&T portfolio as necessary to achieve each of the grand-challenge demonstration goals.

Recommendation #5: Improve Aspects of the Air Force S&T Management Process.

- Obtain HAF-level endorsement of an AFRL planning construct for S&T to provide the stability needed for effective mid- and long-range development of technologies.

- Define and implement a formal process for obtaining high-level inputs from MAJCOMs and product centers in periodic adjustments within the AFRL S&T planning construct.

- Develop and implement an informal process to obtain more frequent inputs from MAJCOMs and product centers for lower levels of the AFRL S&T planning construct.

Chapter 1

Introduction

Technology Horizons was conducted over a nine-month period to provide a vision for science and technology (S&T) in the Air Force over the next decade and beyond. This chapter summarizes the goals, organization, and execution of this effort and describes how the results are presented.

The US Air Force today finds itself at an undeniably pivotal time in its history. It is without question the most effective and powerful air force in the world, and the only air force that can truly project global power. That position of strength was attained by organizing, training, and equipping a professional workforce for the entire range of functions essential to joint combat and combat support, as well as for noncombat missions that the service is increasingly called on to perform. Yet perhaps the single most important factor in achieving this position has been the unmatched technological advantage that the US Air Force has attained over any of its competitors.

Today, however, the confluence of strategic changes, worldwide technological advancements, and looming resource constraints cause some to wonder how the Air Force will maintain this position of superiority. As the spectrum of conflict has grown, so too have the demands for a wider range of capabilities across all facets of conflict. Unprecedented worldwide diffusion of technologies is giving competitors access to capabilities that may have the potential to offset important parts of USAF technological advantages. At the same time, indications are that defense budgets in the coming decade may make it increasingly difficult to achieve the pace of S&T investment that has provided the foundation for US Air Force superiority.

A Vision for Air Force Science and Technology 2010–30

It is against this background that the secretary of the Air Force and the Air Force chief of staff called for the chief scientist of the Air Force to conduct the *Technology Horizons* study (see app. B). The study examines technologies across the air, space, and cyber domains to develop a forward-looking assessment, on a 20-year horizon, of possible offensive and defensive capabilities and countercapabilities of the Air Force and its potential future adversaries that could substantially alter future war-fighting environments and affect future US joint capability dominance. In doing so, *Technology Horizons* provides a vision for where Air Force S&T should be focused over the next decade to maximize the technology superiority of the US Air Force.

A Rational Technical Assessment, Not a Forecast or Prediction

Technology Horizons is neither a prediction of the future nor a forecast of a set of likely future scenarios. Such efforts play a useful role, but forecasts or predictions tend in most cases to overestimate the pace of progress, even when technology overall is advancing at an undeniably rapid pace. Equally important, efforts to forecast or predict the future invariably miss one or more trivial events that can have a determinative effect on the future that actually occurs. Chaos theory indicates that the particular future that will occur beyond a short time horizon is the result of a confluence of many events, the product of which can be inherently unpredictable with any useful certainty and can be influenced to leading order by seemingly incidental events.

> Our Air Force is at another inflection point in its history, where changes in the strategic environment, new technologies, and changes in resources combine to reshape our capabilities and to set us in new directions.
>
> —The Honorable Michael B. Donley
> Secretary of the Air Force

Neither is *Technology Horizons* a fantasy from technologists describing a future in which technology will achieve seemingly boundless wonders. The inaccuracy of such descriptions has over time served more to reveal how difficult it in fact is to translate technological possibilities into practically achievable systems while satisfying numerous operational constraints that may have nothing to do with the technologies themselves.

Rather, *Technology Horizons* is a rational assessment of what is credibly achievable from a technical perspective to give the Air Force a set of capabilities that is suited for the strategic environment of 2010–30. It identifies the key technology areas (KTA) where Air Force S&T investment is most critical for providing the service with the ability to adapt and prevail across the spectrum of conflict in the air, space, and cyber domains. It further addresses emerging cross-domain areas where entirely new threats and opportunities exist.

While this study is visionary, its view is informed by the strategic context in which these technology-derived capabilities will be used. It is an articulation of the "art of the possible" but is grounded in the scientific and engineering knowledge that merely being possible is only a prerequisite to being practically useful. It not only considers the spectrum of technical possibilities but also acknowledges that budget constraints will limit the set of these that can be pursued.

Science and Technology for Materiel and Nonmateriel Solutions

Air Force capabilities and costs are closely interdependent. The costs of normal operations consume resources that otherwise might be available to develop greater capabilities. Manpower costs, for example, are consuming an increasingly larger fraction of the overall Air Force budget. Technology solutions that can reduce manpower requirements or increase manpower efficiencies via nonmateriel means can thus have as much impact on future capabilities as can direct materiel solutions. The S&T vision articulated in *Technology Horizons* thus seeks to identify "disproportionately valuable" materiel and nonmateriel solutions over the 2010–30 time frame.

Technology Horizons looks at capabilities on a 20-year time horizon but recognizes that it can take a decade to translate technologies from maturity into fielded systems. It therefore is based on an approach that considers technology developments over

> *While we remain resolute about the issues that remain, we can, and we must, raise our sights to focus on the longer-term vision—an Airman's vision of constant innovation in the control and exploitation of air, space, and cyberspace.*
>
> —Gen Norton B. Schwartz
> US Air Force Chief of Staff

the next decade that can provide the basis for technology-enabled capabilities over the following decade.

This effort also recognizes that increasingly more of the S&T that provides the basis for future Air Force capabilities is available worldwide to be translated into potential adversary capabilities. It consequently has sought to envision not only US joint and allied opportunities for using these technologies, but also ways that adversary capabilities and "technology surprise" could be derived from them using entirely different concepts of operations or on the basis of entirely different warfighting constructs.

Technology Horizons further recognizes that capabilities enabled by new technologies and associated operating concepts often introduce new vulnerabilities not envisioned in the original capability. It thus has sought to consider potential vulnerabilities and cross-domain interdependences that may conceivably be created by second- and third-order effects of these technology-derived capabilities.

Can We Even Afford to Do Any of These Things?

With factors such as manpower, sustainment, and energy costs consuming increasingly more of the Air Force budget, and the outlook for defense budgets suggesting little or no real growth over the next decade, some may rightly ask whether a study such as this is even useful. If the Air Force does not have the resources to put behind the S&T focus areas emerging from this effort, then what sense does it make to identify these and develop an Air Force S&T vision around them?

At a minimum, such a view would overlook the fact that technologies themselves can help offset many of the costs noted above. For instance:

- New technologies for increased cyber resilience of Air Force networks and systems can potentially free up manpower otherwise consumed by additional cyber specialists and their training that would be needed to defend these systems.

- Technologies can provide increased trust in autonomy to enable reduced manpower requirements via flexibly autonomous systems, or equivalently autonomous systems can enable greater war-fighter capabilities for the same manpower requirements.

- Technologies can enable fuel cost savings by increases in turbine engine efficiency, advances in lightweight materials and multifunctional structures, advanced aerodynamic concepts and technologies, and adaptive control technologies.

- Technologies could potentially reduce manpower needs through augmentation of human performance via implants that improve memory, alertness, cognition, and visual/aural acuity; brainwave-coupled human-machine pairings; or even screening of individuals for specialty codes based on brainwave patterns or genetic correlators.

Such examples show how an Air Force S&T vision that properly accounts for the role that budget constraints will play in the strategic environment can enable materiel and nonmateriel solutions to reduce Air Force costs while expanding capabilities.

It should be further apparent that in a budget-constrained environment it is all the more essential to have a clear vision for where S&T investments should be focused. It is precisely when resources are too constrained to allow as broad a set of S&T investments as might be desirable that the Air Force needs a vision for which technology areas are most essential for it to invest in. While broader investment will be needed to provide an appropriately hedged strategy, a substantial fraction of total S&T efforts should be aligned with a vision that identifies technology areas for enabling among the greatest Air Force capabilities.

> Technology Horizons *provides a vision for where a substantial fraction of Air Force S&T investment should be focused to provide the greatest possible capabilities in the strategic environment, budget environment, and technology environment during 2010–30.*

Lessons Learned from Prior Air Force Science and Technology Visions

Throughout its history, beginning shortly before it became a separate service, the Air Force has conducted a major study roughly once every 10 years to develop a vision for the role that S&T would seek to fill over the following decade. As shown by the timeline in figure 1, six such studies have previously been conducted, beginning with *Toward New Horizons* in 1945 led by Theodore von Kármán for Gen Hap Arnold, and continuing through *New World Vistas* conducted in 1995 by the Scientific Advisory Board for secretary of the Air Force Dr. Sheila Widnall and Air Force chief of staff Gen Ronald Fogleman.

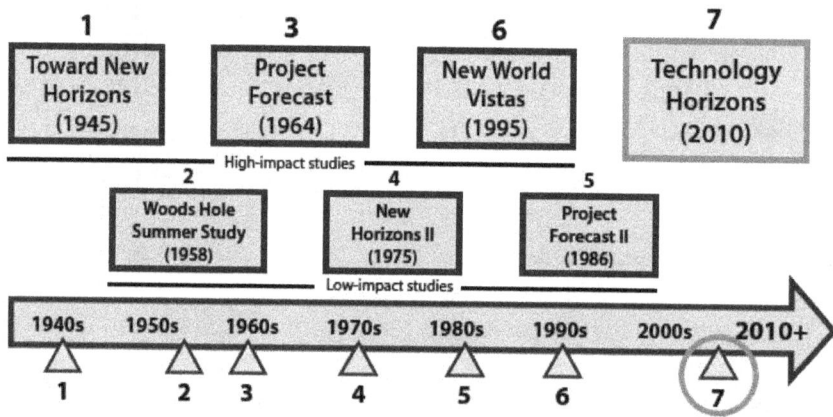

Figure 1. Summary of Headquarters Air Force–level studies to produce Air Force S&T visions. *Technology Horizons* is an overdue effort to produce an S&T vision appropriate for the strategic environment during 2010–30. (Courtesy of the Office of the Chief Scientist of the Air Force.)

Since completion of the latter, over 15 years have passed without a major study to develop an updated S&T vision for the Air Force. *Technology Horizons* is the next in this succession of major S&T visions conducted at the Headquarters Air Force (HAF) level. Particularly in view of the fundamental strategic changes, rapid global technological advancements, and forseeable resource constraints that define the environment over the next decade, *Technology Horizons* is an overdue effort to help guide Air Force S&T investments to include areas that will have disproportionately valuable impacts on reducing costs and maintaining capability advantage.

> Technology Horizons *seeks to help guide the S&T efforts of the Air Force over the next decade to maximize their impact in the substantially different environment that the Air Force is now facing.*

Previous Air Force Science and Technology Visions

Figure 1 shows the six previous Air Force S&T vision studies that have been conducted at the HAF level. The duration, size, composition, organization, output, and impact of these have varied widely. The following observations can be made:

- These studies have been conducted over durations ranging from nine months to, in one case, nearly two years.

- The numbers of participants involved has ranged from as few as 25 to nearly 500.

- Contributors have ranged from academia to the military, the government, and industry.

- Panels have in some cases been organized around technical disciplines, and in others around capability topics, as well as other arrangements.

- Reports have ranged from concise summaries to more than a dozen volumes and total lengths in excess of 1,300 pages.

- Impacts on technical directions, organization, and other aspects of Air Force S&T have ranged widely, in part depending on how responsive the studies were to the central issues of their time.

Michael Gorn gives detailed descriptions of the first five of these prior Air Force S&T vision studies in his book *Harnessing the Genie: Science and Technology Forecasting for the Air Force 1944–1986*.

Key Factors That Have Produced Successful S&T Visions

As the above summary suggests, and as detailed in the reference noted, there is no correlation of the impact of these studies with either the number of participants or the length of the reports they have produced. What correlations there are have principally to do with the types of participants and the way that working panels have been organized.

Studies without representation in substantial numbers from the research community have generally not been successful, though sufficient operational inputs are essential to ground the discussions. Panels organized along relatively narrow system themes appear, as might be expected, to restrict the range of ideas for discussion, and essentially preclude cross-domain insights created by interdependences among systems. Studies with panels organized around broader themes have generally proven to be more successful.

> *The principal factor that has determined the impact of prior S&T visions has been the extent to which Air Force senior leadership has embraced them as a means for helping to vector the direction of future Air Force research.*

Organization and Conduct of *Technology Horizons*

Technology Horizons formally began in June 2009 and was completed in February 2010. Organization of the study and the working groups preceded the formal start of the effort.

Strategic Perspective as the Foundation for an Effective S&T Vision

The Air Force S&T vision from *Technology Horizons* is predicated in part on the principle that S&T can only be usefully guided in a study like this by placing it in context with the strategic environment in which the Air Force will be operating during the 2010–30 time frame it addresses. The technology areas that are recommended for Air Force S&T to put special emphasis on are those that meet key needs dictated by this strategic environment. *Technology Horizons* thus is not simply an opportunity-driven technology vision but matches the demand-side pull of the Air Force during 2010–30 with the opportunity-side push that realistically achievable technologies can enable.

Technology Horizons "10+10 Technology-to-Capability" Process

To match technology opportunities with the needs of this strategic environment, the study used a 10+10 Technology-to-Capability process, shown in figure 2.

This recognizes that even if it were possible to project the state of *technologies* 20 years into the future to the 2030 horizon date of the study, this effort is instead designed to help the Air Force envision what technology-enabled *capabilities* it could credibly have to meet its key

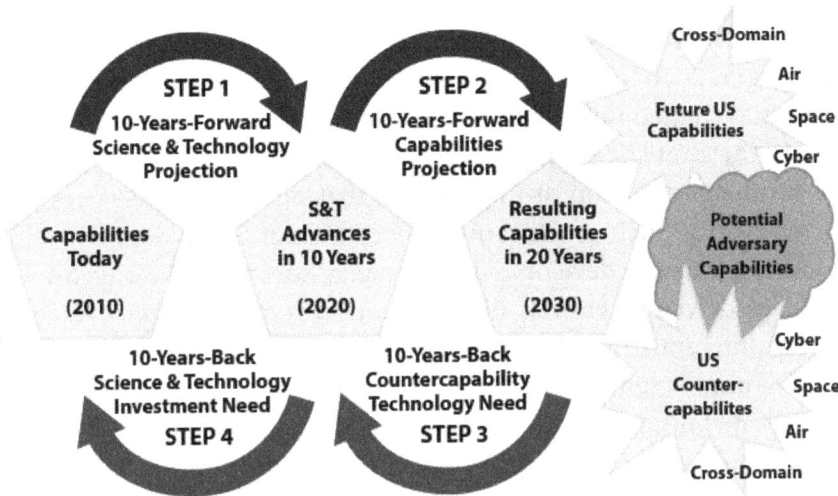

Figure 2. Schematic of the 10+10 Technology-to-Capability process used in *Technology Horizons*. (Courtesy of the Office of the Chief Scientist of the Air Force.)

needs in 2030. Thus, the study must determine the state to which the necessary supporting technologies can be brought in time to enable those capabilities by 2030.

Owing to the 10 years or so that it typically takes to transition technology readiness level (TRL) 6 technologies into fielded capabilities, this means that the underlying date to which technologies must be projected is 2020. That, in turn, allows a determination of the technology investments that need to be under way today to enable these Air Force capabilities in 2030.

The required 10-year projection from today's state of technology can be credibly made with acceptable uncertainties, even under the rapid pace at which technologies are advancing. By contrast, the uncertainties if a 20-year projection of technologies were needed

Technology Horizons will help the Air Force obtain the greatest effectiveness in a budget-constrained environment; if we invest in the right technology areas we can have unbeatable capabilities.

from today's state would be far greater than twice as large. Thus, the 2030 time horizon of *Technology Horizons* requires only a credible projection from 2010 to 2020 of the technologies needed to enable a set of capabilities that could be fielded in 2030 to meet key needs of the Air Force in the strategic environment.

A further foundation of this study is that potential adversaries will have access to much of the same S&T as the Air Force does over this period and thus can develop red force capabilities from these technologies that—driven by entirely different concepts of operations—may be entirely different from blue force capabilities developed from them. It thus becomes essential to envision not only credible US technology-derived capabilities in 2030, as shown on the right in figure 2, but also potential adversary capabilities along with appropriate US counter-capabilities, as also shown on the right in the figure.

Developing the technologies needed to enable these counter-capabilities then involves another pair of 10-year steps, in this case shown in red in figure 2. The first of these steps begins with the needed countercapabilities in 2030 and projects back to the required state of the underlying enabling technologies in 2020. The second of these then does a further 10-year projection back to 2010 to determine what S&T investments would need to be under way today to thereby enable these countercapabilities in 2030.

The result is the four-step 10+10 Technology-to-Capability process in figure 2:

- Step 1: Beginning with the state of technologies today, a 10-year forward projection is made of the state to which these technologies, or ones derived from them, can be brought in 2020.

- Step 2: From the state to which technologies can be brought in 2020, a further 10-year forward assessment is made of the US capabilities that could be enabled by them to meet key needs of the strategic environment in 2030.

- Step 3: Potential adversary capabilities and needed US counter-capabilities are envisioned in 2030, and a 10-year backward projection is made of the state to which technologies needed to enable these countercapabilities must be brought by 2020.

- Step 4: From the required state of technologies for US capabilities and countercapabilities in 2020, a further 10-year backward projection is made to identify the needed S&T activities that must be under way today to enable these.

In practice the precise dates must necessarily be interpreted somewhat notionally. In the cyber domain, for instance, it is essentially impossible to project technology 10 years into the future, and in that case the process was instead applied in a "5+5" framework.

Similarly, while this process has formed the foundation of the study, its implementation has been as much qualitative as quantitative. A fully quantitative assessment for each of the technologies considered here would have been beyond the scope even of this effort. Moreover, by ensuring that working groups and other sources of inputs to this effort included subject matter experts with technical and operational knowledge, quantitative aspects of this process could be addressed in part by assessments of these 10-year projections based on technical and operational judgment.

Identifying technology focus areas that will be most essential over this period has as much to do with technical and operational judgment as with direct quantitative analysis. As with any vision, the nature of S&T forecasting prevents it from being reducible to an algorithm that might seek to obtain similar insights. An articulation of the S&T that will be most valuable for the Air Force over the next two decades is ultimately the result of a set of reasoned, technically rooted, and objective judgments informed by an understanding of their strategic context.

Broad Range of Inputs to *Technology Horizons*

As figure 3 shows, *Technology Horizons* consists of five phases, including three working phases designed to obtain a broad range of subject matter expertise in the implementation of the underlying process summarized above. Phases 2 and 3 included working groups, one each in the air, space, and cyber domains, and a fourth group that focused on cross-domain insights. The functions of these groups, their compositions, and their participants' backgrounds are summarized in appendix E. The working groups included representation from the

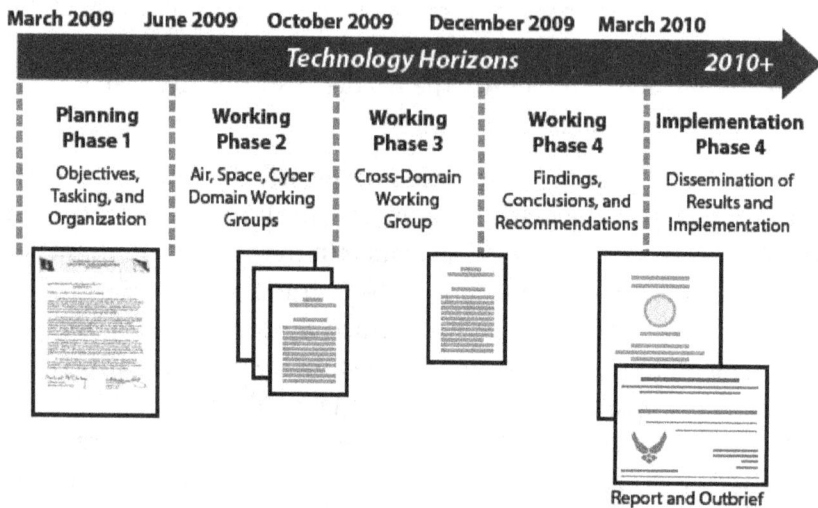

| March 2009 | June 2009 | October 2009 | December 2009 | March 2010 |

Technology Horizons · 2010+

Planning Phase 1	Working Phase 2	Working Phase 3	Working Phase 4	Implementation Phase 4
Objectives, Tasking, and Organization	Air, Space, Cyber Domain Working Groups	Cross-Domain Working Group	Findings, Conclusions, and Recommendations	Dissemination of Results and Implementation

Report and Outbrief

Figure 3. Organization of *Technology Horizons* into planning, working, and implementation phases to provide an S&T vision that supports Air Force capability needs in the 2010–30 strategic environment. (Courtesy of the Office of the Chief Scientist of the Air Force.)

- Air Force S&T community,

- intelligence community,

- MAJCOMs,

- product centers,

- federally funded research and development centers (FFRDC),

- defense industry, and

- academia.

In addition to inputs from these groups, *Technology Horizons* obtained a broad range of further inputs from site visits, briefings, and discussions involving organizations across the Air Force and elsewhere in the Department of Defense (DOD), federal agencies, FFRDCs, national laboratories, and industry. These included inputs from Air Staff and Air Force secretariat offices, MAJCOMs and product centers, direct reporting units, and field units. A list of these additional sources of inputs is given in the bibliography.

Additional operational perspectives were gained from briefings, visits, and discussions with

- Air Combat Command (ACC),

- Air Force Special Operations Command (AFSOC),

- Air Force Space Command (AFSPC), and

- Air Mobility Command (AMC).

The effort further made use of a wide range of perspectives in nearly 200 reports and technical papers from the Air Force Scientific Advisory Board, the Defense Science Board, and other organizations. A partial list of these additional sources is given in the bibliography.

> This exceedingly broad range of inputs to Technology Horizons has allowed development of a forward-looking, balanced, and effective S&T vision for the Air Force in 2010–30.

A Focus on Cross-Domain Insights

While the working group phases were organized along the air, space, and cyber domains, it was recognized that some of the most important results from *Technology Horizons* would deal with cross-domain insights. A further working group was thus formed to address cross-domain effects, defined here to include the following:

- A technology in one domain that can produce unexpected effects in another domain; the effects could be beneficial or detrimental.

- A technology in one domain that requires supporting functions in another domain and thus creates interdependencies between two or more domains.

- A technology that falls "between" the classical domains but has implications in one or more of them.

Given the emphasis on cross-domain insights in this effort, the results from *Technology Horizons* are intentionally not presented along traditional domain boundaries.

> *The understanding and exploitation of cross-domain effects will grow in importance in the next decade and beyond, and the organization and execution of Air Force S&T will need to ensure access to such cross-domain insights and capabilities.*

Organization of Results from *Technology Horizons*

Strategic Context

Enduring Realities

Overarching Themes

Potential Capabilities

Key Technology

Grand Challenges

Principal Findings

Implementation Plan

Figure 4. Major elements of the *Technology Horizons* **vision for Air Force S&T over the next decade.** (Courtesy of the Office of the Chief Scientist of the Air Force.)

Chapter 1: Introduction. Summarizes the Air Force vision for S&T, lessons learned from its prior S&T visions, organization and conduct of the study, organization of results, and caveats for the study.

Chapter 2: Strategic Context for Air Force S&T 2010–30. Summarizes the principal strategic factors that will drive needed Air Force capabilities during this time and factors relevant to the worldwide S&T arena that will impact the Air Force's ability to maintain superior technological capabilities over this time frame.

Chapter 3: Enduring Realities for the Air Force 2010–30. Recognizes key drivers that will remain largely unchanged during this time and that will constrain the Air Force's ability to shape itself for the future using S&T in ways it might in an otherwise unconstrained environment.

Chapter 4: Overarching Themes for Air Force S&T 2010–30. Identifies specific overarching themes that will be central for meeting Air Force needs dictated by the above strategic context and enduring realities; these themes form the essential foundation for the most important Air Force S&T during 2010–30.

Chapter 5: Technology-Enabled Capabilities for the Air Force 2010–30. Defines a set of technologically achievable capability areas that are well aligned with the overarching themes identified above and then uses these notional capabilities to identify key enabling technologies that are most impactful across this set.

Chapter 6: Key Technology Areas 2010–30. Determines a cross-cutting set of enabling technology areas that are most determinative over the notional capabilities identified above and identifies them as being among the most important areas to emphasize in Air Force S&T over the next decade and beyond.

Chapter 7: Grand Challenges for Air Force S&T 2010–30. Lists a set of challenge problems that will help focus Air Force S&T over the next decade on the KTAs identified above through technology development efforts followed by systems-level integrated technology demonstrations to achieve stretch goals.

Chapter 8: Summary of *Technology Horizons* Vision. Describes the major S&T focus areas that the Air Force should emphasize during the next decade and beyond to enable technologically achievable capabilities that will give it the greatest joint force effectiveness by 2030.

Chapter 9: Implementation Plan and Recommendations. Outlines the proposed plan for implementing the S&T vision from *Technology Horizons*, provides actionable recommendations to vector Air Force S&T over the next decade, and identifies corresponding primary and supporting organizations to implement these.

> *The above chapters of* Technology Horizons *give the results of a rational approach for determining—from the enormous range of technologies that could benefit the Air Force—a set of guiding principles for the most important S&T to be conducted over the next decade.*

Caveats

Value of Technologies Not Specified in *Technology Horizons*

No effort such as this can usefully list all S&T that is important for the Air Force to pursue, nor has it been an objective of *Technology Horizons* to do so. Instead, this effort has sought to identify some of the most valuable technology areas for Air Force S&T—ones that the Air Force must pursue to enable key capabilities that it will need to be as effective as possible in the strategic environment during 2010–30. The KTAs identified herein thus necessarily represent only the most essential fraction of the overall research portfolio that should be pursued by the Air Force.

Focus on Key Technology Areas versus Broad Technical Domain Areas

Technology Horizons was designed to identify KTAs that are among the most essential for Air Force S&T to focus on over the next decade and beyond. These technology areas aggregate individual narrow research topics into usefully defined focus areas that Air Force S&T should emphasize but avoid dictating the individual lines of research within these areas that might prove to be most productive. Focusing on such KTAs is the appropriate level of specificity for an S&T vision.

At the other extreme, overly broad domain area descriptors such as "material science," or even broad technical domains within these such as "nanotechnologies," do not provide a comparable level of specificity needed to usefully serve as a guide for Air Force S&T. KTAs identified as most essential for the Air Force to pursue are sufficiently specific to usefully guide S&T investment choices but are not so narrow as to dictate individual research projects to be pursued.

Connection between Air Force S&T and the Acquisition Process

Air Force S&T is the initial phase of the process by which technologies are matured and, where appropriate, are transitioned for acquisition by the Air Force. As a consequence, often-discussed improvements to the defense acquisition process could potentially improve the transition of technologies from S&T into systems for acquisition. At least some of the elements of the S&T vision identified in *Technology Horizons* could potentially achieve greater likelihood of practical impact from improvements to the defense acquisition process. However, attempts to suggest such improvements to the acquisition process were outside the scope of this effort.

Chapter 2

Strategic Context for
Air Force S&T 2010–30

Air Force S&T can be prudently vectored only when placed in the context of the strategic environment in which the Air Force will be operating during this period. This chapter summarizes key elements of this environment that will drive needed Air Force capabilities over the next two decades.

The strategic environment that the Air Force faces over the next two decades is substantially different from that which has dominated throughout most of its history. Since the end of the Cold War, long-standing paradigms that had ordered thinking and policy with respect to conflict have been replaced by a far broader range of threats and a less predictable set of challenges. Potential conflict scenarios now include not only major powers and regional players, but also failed and failing states, radical extremists, and a range of nonstate actors that spans from organized militias, informal paramilitary organizations, warlords, and warring ethnic groups to pirates, organized criminals, terrorists, and even individuals who may come from around the globe or across the street. Potential drivers of conflict range from religious extremism and ethnic disputes to resurging nationalism and aspirations for regional influence, and from competition over energy and natural resources to the need to contain nuclear proliferation and the spread of chemical, biological, and radiological weapons.

Unable to compete in direct engagements with US joint forces, adversaries have learned to exploit newfound asymmetric advantages. Some have sought to shift the battle into the media, exploiting public intolerance for real or contrived collateral damage—strikes that by historical standards are being conducted with near surgical accuracy are now often deemed unacceptable. Others have learned to exploit US dependence on the space and cyber domains, developing ways to disrupt or deny these to achieve potentially far-reaching cross-domain effects. Some are seeking access to nuclear capabilities. The Cold War, by comparison, seemed remarkably simpler, and the set of capabilities that the Air Force needs to be effective in the foreseeable strategic landscape is substantially different and appears less certain than ever before.

To aid in deterring conflict with major world powers, the United States must correctly understand their potential capabilities in the 2030 time frame, many of which are derived from investments in technologies that support development of military systems. China, for example, has focused much of its recent military modernization on investments in high-end asymmetric capabilities, emphasizing electronic warfare (EW), cyber warfare, counterspace operations, ballistic and cruise missiles, advanced integrated air defense systems (IADS), and theater unmanned air vehicles. By combining imported technologies, reverse engineering, and indigenous development, it is seeking to rapidly narrow the technology and capability gap between the People's Liberation Army Air Force (PLAAF) and the US Air Force. China is also using "military diplomacy" to expand security cooperation activities with Asian states and engage foreign militaries in a range of cooperative activities. Russia as well is developing and fielding new technology-derived military systems, including advanced fighter aircraft such as the T-50 PAK FA and advanced IADSs such as the SA-21. Global diffusion of advanced military systems through arms transfers, not only by Russia and China but also by France, Israel, and many others, is making substantial technology-derived capabilities available around the world.

Among these, China in particular is widely regarded as having both the economic resources to devote to such advanced technology-development programs and the national desire to achieve the resulting regional and global influence that such systems may bring. Its growing economy has helped pay for a massive military modernization program that includes not only new fighter aircraft but also naval vessels and missiles.

Yet at the same time, China's one-child policy has created serious demographic imbalances in gender and age distributions, which are requiring it to direct resources toward social programs that could potentially constrain these ambitions. As a result, carefully chosen cost-imposing strategies developed through appropriate US technology-enabled capabilities can provide an opportunity to slow the potential threat that this military buildup creates. If we understand the strategic environment that the Air Force faces over the next two decades, we can make the right choices to obtain the mix of capabilities that will best meet US national security objectives.

Relation to National Security Objectives

Military forces exist to develop and offer a range of operational and strategic options to the president for meeting national security objectives and to joint force commanders for meeting military objectives. Throughout its history, the US military's consistent dual purpose has been not only to fight and win the nation's wars but also to protect the nation and its global interests in ways that extend beyond direct combat operations. The latter acknowledges that military forces are as much an instrument employed for shaping the global environment, for deterring those who might otherwise harm the nation's interests, and for providing regional stability where needed to advance the nation's broader interests as they are for warfare itself.

At the broadest level these functions include the following:

- Protecting and defending the homeland from external attack.

- Deterring conflict with major global powers and encouraging regional stability through use of Air Force global vigilance, global reach, and global power.

Figure 5. Supporting US joint force operations in so-called irregular or hybrid warfare while preparing for possible larger conflicts with a near-peer adversary is among the Air Force's greatest challenges. (Courtesy US Air Force.)

- Preparing to fight and win the nation's wars as part of the joint force.

- Providing military assistance to civilian authorities.

- Enabling national and partner instruments of power to prevent or contain local or regional instabilities.

- Supporting the nation's cooperative relationships with international partners and its interactions with competitors.

In fulfilling these objectives to protect the nation and advance its global interests in times of peace, crisis, and combat, the Air Force is distinct among the joint services in its ability to use the speed, range, flexibility, precision, and lethality of aerospace forces to provide global vigilance, global reach, and global power. Its mission requires it to look at the world from an inherently strategic perspective as it conducts global, regional, and tactical operations in the domains of air, space, and cyberspace in concert with the other services, with other national instruments of power, and with US international partners. This requires capabilities for the full spectrum of nonconflict and conflict operations, ranging from emergency response and humanitarian relief to counterinsurgency operations, major warfare, and homeland defense.

Technology-Derived Challenges to Air Force Capabilities

The immense asymmetric advantage that the Air Force's air dominance has for much of the past 50 years provided for US and partner forces could be potentially put at risk by worldwide development and proliferation of numerous advanced-technology-derived threats, including integrated air defenses, long-range ballistic missiles, and advanced air combat capabilities. There have been equally important advances in counterspace technologies, in cyber warfare technologies, and in understanding the cross-domain effects that these technologies can produce on the US ability to conduct effective air, space, and cyber domain operations. The combined effects of these and yet further technology advances that will occur over the next decade, and their transition into worldwide military systems, are essential elements of the stra-

Figure 6. High-end systems such as the F-22 support Air Force air dominance over the near term to midterm. *Technology Horizons* seeks to define the S&T that will enable the next generation of Air Force capabilities that are suited to the needs of the 2010–30 strategic environment. (Courtesy US Air Force.)

tegic context that determines the most essential Air Force S&T over this period.

Advanced Electronic Systems

Land-based air defenses and air combat systems in particular are benefitting from increasingly widespread use of advanced electronics, including active electronically scanned array (AESA) radar technologies to enable more capable detection and targeting capabilities. These have evolved to counter nontraditional threats such as remotely piloted aircraft (RPA) and cruise missiles. Advanced EW systems, including approaches based on digital radio frequency memory (DRFM) sys-

Figure 7. Advanced fighter development efforts, such as the fifth-generation Russian/Indian T-50 PAK FA, could pose a significant mid- to long-term challenge to US joint air dominance. (Reprinted from Wikipedia, s.v. "Sukhoi PAK FA," accessed 20 June 2011, http://en.wikipedia.org /wiki/Sukhoi_PAK_FA.)

tems, that are being developed and proliferated today have shown significant abilities even against advanced AESA radars. Further advances in miniaturization and speed of electronics—and the massive processing capabilities that these are enabling—are likely to yield further substantial increases in the capabilities of EW systems over the next decade and beyond.

Advanced Missile Seekers

Advances in electro-optical tracking, such as dual-band electro-optical-infrared (EO-IR) imaging arrays with wide fields of view and advanced digital tracking filters, can be expected to significantly increase over the next decade. Early implementations of combined EO-IR imaging seekers in systems such as the highly maneuverable beyond-visual-range Python 5 air-to-air missile (AAM) have shown the effectiveness of these technologies. Corresponding implementations of these advanced technologies in surface-to-air missiles (SAM), such as the SPYDER antiaircraft missile system based on the Python 5, are extending these technology advances to area defense systems. Such lethal systems are already beginning to be proliferated via international sales. Their capabilities will appear even more broadly over time, both through indigenous developments that use these technologies and through subsequent global sales of such systems.

High-Speed Air-Breathing Missile Propulsion

High-speed air-breathing missile propulsion technologies have steadily advanced over the past two decades and are now enabling substantially more capable supersonic AAMs, SAMs, and cruise missiles. Such a ramjet-derived propulsion system is used in the MBDA-developed Meteor, a Mach 4+ AAM that is likely to enter service in 2013. A ramjet-derivative propulsion system is also being used in the Russian/Indian BrahMos Mach 2.5+ cruise missile, brought into service in 2007, and a related ramjet-based air-breathing propulsion system is used in the Indian Mach 2.8+ Akash SAM, first fielded in 2009. The speed, range, and terminal maneuverability benefits of such air-breathing supersonic propulsion technologies will likely accelerate development of missile systems to exploit these advantages.

Advanced Integrated Air Defense Systems

Development and proliferation of advanced IADSs have progressed substantially. Longer-range mobile SAMs in the Russian S-300 family, such as SA-10s/20s, which have been exported widely, and corresponding Chinese-manufactured derivatives of "double-digit SAMS," such as the HQ-10/15/18, represent significant challenges. These systems can potentially engage legacy fighters at ranges beyond their own ability to hold such targets at risk. More advanced Russian systems, such as the S-400, have advanced tracking and longer-range capabilities and could potentially be sold internationally. Modern integrated systems such as these are enabling a fundamental change in air defense strategy, from traditional point defense of key targets to broader antiaccess/area-denial (A2/AD) approaches based on offensive and defensive counterair operations.

Passive Sensors and Electronic Warfare

Further development of passive sensor technologies such as infrared search and track (IRST) systems over the next decade or more will make air defense systems increasingly resistant to electronic suppression. Potential adversaries are also upgrading to networked air defense systems with advanced electronics and signal processing capabilities that can make jamming far more difficult. AESA radars are allowing active electronic beam forming and steering. EW capabilities of potential adversaries are increasingly shifting from jamming to sophisticated spoofing, made possible largely by the continuing miniaturization of commercial electronics. Availability of inexpensive yet massively capable electronic processing and storage not only allows the number and sophistication of such devices to grow rapidly but also to be fielded in relatively low-cost EW pods and decoys. Software changes allow these threat systems to be readily adapted to new countermeasures.

Figure 8. Technologies such as the ramjet-derivative propulsion system on the MDNA Meteor are beginning to enable high-speed AAMs with greater range and terminal maneuverability. (Courtesy MBDA–Th Wurtz.)

At the same time, an overlap has developed between electronic and cyber warfare, especially in the wireless radio-frequency (RF) domain. The technological sophistication, diversity, and proliferation of advanced EW systems present a substantial challenge to US air dominance. New technology-derived approaches will be needed across the spectrum of standoff and close-in EW capabilities, including electronic attack, protection, and support techniques suited to permissive, contested, and highly contested environments.

Shoulder-Fired Surface-to-Air Missiles

Next-generation shoulder-launched SAM systems can be expected to further add to the threats that advanced technologies pose. Current "low end" man-portable air defense systems (MANPADS) are affordable enough to buy in large quantities and can be readily dispersed, hidden, and employed for close-in air defense. Further advances in miniaturization of electronics and processing are likely to produce substantially more capable systems at price points that may permit comparably widespread global diffusion.

Ballistic and Cruise Missiles

The emergent long-range air defense capabilities of several potential adversaries have serious implications for US airpower. Additionally, ballistic and cruise missiles with growing mobility, range, maneuverability, and precision pose further threats to Air Force systems and their ability to deliver the power projection on which much of US strategy is based. This is especially the case in the western Pacific, where China has fielded conventional ballistic missiles and cruise missiles with significant reach. By 2015 China is expected to have hundreds of DF-15 ballistic missiles and DH-10 cruise missiles capable of reaching much of the western Pacific. Advanced medium-range ballistic missiles such as the DF-21, DF-25, and the intercontinental-range DF-31 may be in service in significant numbers, and a DF-41 advanced intercontinental ballistic missile (ICBM) may be under development.

The threats posed to US air bases in the Pacific by ballistic missiles create significant technology challenges for maintaining airpower projection in this region. The long ranges required for operations in the

Pacific entail substantial tanking requirements that put a premium on long-range strike capabilities and fuel-efficient propulsion systems. The air-sea battle concept based on synchronized Air Force and Navy operations against a potential near-peer competitor in this region may also require significant new capabilities.

Figure 9. Dual-mode EO-IR focal-plane array seeker technology, as on the Python-5 AAM, can give wide field-of-view imaging infrared capability and full-sphere beyond-visual-range engagement. (Reprinted from *Wikipedia*, s.v. "Python [missile]," accessed 20 June 2011, http://en.wikipedia.org/wiki/Python_(missile).)

Figure 10. Advanced SAM systems, such as the SPYDER that combines Python-5 EO-IR missiles with active radar-guided missiles, can provide highly capable area defense capabilities. (Reprinted from *Wikipedia*, s.v. "SPYDER," accessed 20 June 2011, http://en.wikipedia.org/wiki/SPYDER.)

Advanced Fighter Aircraft

Russian development of the Sukhoi T-50 PAK FA fifth-generation fighter, reportedly in a joint effort with India, is an indicator that the United States must continue to develop its capabilities to maintain air dominance. The aircraft will reputedly include substantial technology-derived advances over all prior potential competitor aircraft, including significant low observable (LO) capability, advanced AESA radar, advanced avionics, and supercruise capability. Early flights of the PAK FA in January and February 2010 suggest a serious development program that has reached an advanced stage of engineering refinement. Reported Chinese efforts to develop one or more J-XX fifth-generation fighters appear not to be as far advanced, but the resources that may be available to devote to these efforts suggest such a capability could be fielded within the time horizon of this study. In general, fifth-generation fighter systems, if deployed against the United States, have the potential to pose substantial challenges to our air dominance over the next decade.

Remotely Piloted Aircraft

Growth in military use of remotely piloted vehicles has been rapid as forces around the world explore increasingly wider uses for them, including surveillance, strike, EW, and others. These will include fixed-wing and rotary-wing systems, airships, hybrid aircraft, and other approaches. They will have increasingly autonomous capabilities allowing remote pilots to declare their overall mission intent but permit these systems to adapt autonomously in the local environment to best meet those objectives. Some systems may operate collaboratively in multiple-craft missions to increase survivability and deliver effects that could not be achieved individually. Technologies are enabling such increasingly autonomous systems to exchange data on their respective states in order to adjust their mission planning and adapt to their changing environment. Autonomous aerial refueling capability will enable long-range and long-endurance operations. While most remotely piloted systems will likely have limited capabilities, some will incorporate LO technologies, advanced EW functions, and directed energy (DE) strike capabilities. Price points of some systems could allow them to be acquired in numbers sufficient to present significant challenges for certain types of missions.

Directed-Energy Systems

Laser-based and high-power, microwave-based DE systems being developed by several nations will play an important role in the strategic context of 2010–30. The success of MacDonald, Dettwiler and Associates (MDA) in tracking, targeting, and destroying a representative missile in February 2010 with an integrated, airborne, megawatt-class chemical laser demonstrated the potential of strategic-scale systems. Ground-based lasers are likely to appear for air defense and other roles, as will airborne microwave-based systems that can disable or defeat electronic systems. More recent solid-state laser technologies are enabling tactical-scale systems for potentially revolutionary airborne self-defense and low-collateral-damage strike capabilities. Emerging fiber laser technologies as well as diode-pumped alkali lasers may allow later versions of such systems to be made even smaller for integration in a much wider set of platforms, including fighters. DE systems will be among the key "game changing" technology-enabled capabilities that enter service during this time frame.

Figure 11. The stand-up of Air Force Global Strike Command (AFGSC) has consolidated all Air Force assets for the nuclear mission, including B-52s and B-2s, under a single major command. (Courtesy US Air Force.)

Space Control

US joint force dependence on the space domain is a further key component of the strategic context for Air Force S&T during 2010–30. Reliance on assured access to military space systems is essential to meet the demand for global intelligence, surveillance, reconnaissance (ISR), and communications. Potential adversaries recognize our critical dependence on space assets and are developing means for disrupting our access to those assets, or even for incapacitating or destroying those assets. Jammers allow interference with satellite uplinks and downlinks, and cyber attacks can disrupt ground control segments. Some are developing the means to gain access to space themselves. Others are making use of increasingly capable commercial space-based imaging capabilities and space-based imagery freely available via the Internet at levels of resolution that not long ago were accessible only to the intelligence community.

Figure 12. Continuing to strengthen the Air Force nuclear enterprise, such as with the AGM-129A advanced nuclear cruise missile shown here, remains the top US Air Force priority. (Courtesy US Air Force.)

Global Positioning System Denial

The broad dependence of US joint war-fighting capabilities on precision navigation and timing from Global Positioning System (GPS) satellites has made local or regional GPS denial a high priority for adversaries. Beyond preventing access to positioning information, inexpensive low-power GPS jammers and spoofers can deny accurate timing information that is at least equally critical to US systems. China's efforts to field its Beidou-2 (Compass) positioning, navigation, and timing (PNT) satellite network and Russia's fielding of its Global Navigation Satellite System (GLONASS) suggest the role that navigation warfare via GPS denial and spoofing may play in the strategic thinking of these and other nations.

Space Launch Capabilities

Altogether, 11 nations today have orbital launch capability, and roughly 50 possess orbiting satellites. In 2009, with the launch of its *Omid* telecommunications satellite, Iran became the ninth country to have developed the capability to place indigenously developed satellites into low Earth orbit (LEO) using indigenously developed launch systems. This was achieved with its Safir-2 space launch vehicle, a derivative of the Iranian-developed Shahab-3 medium-range ballistic missile (MRBM) that itself is believed to trace to the North Korean Nodong-1 MRBM. A subsequent Iranian launch in February 2010 showed the capability of its Kavoshgar-3 rocket.

Iran further claims to be developing a larger Simorgh orbital launcher, and North Korea too is seeking to develop a satellite launch capability. Countries with these sophisticated capabilities could potentially develop ICBM systems. Moreover, private corporations now operate 19 launch sites throughout the world, providing launch options for nations seeking to gain access to space. In 2009 the first orbital launch using a commercially developed liquid-fueled rocket occurred when Space-X put a Malaysian satellite into LEO using its Falcon-1 launcher, and its evolved expendable launch vehicle (EELV)-class Falcon-9 rocket had its first launch in 2010.

Satellite Technologies

Technologies are also enabling a range of "small sats" with masses below about 200 kilograms (kg), some of which can already today provide significant military capabilities. They include microsatellites with masses between 10 and 100 kg, nanosatellites with masses between 1 and 10 kg, and picosatellites with masses below 1 kg. The latter include "CubeSats" measuring just 10 centimeters on a side, which meet published design standards that allow their deployment by common low-cost mechanisms.

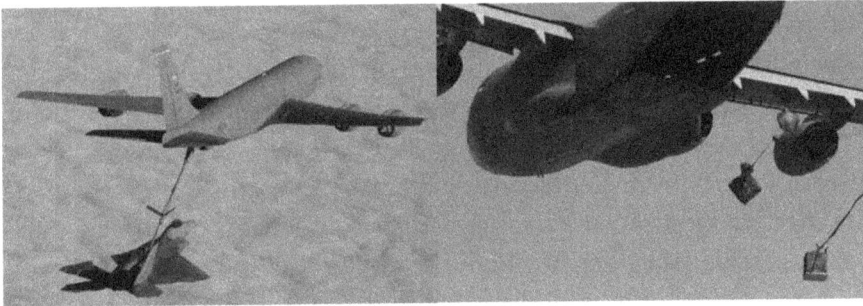

Figure 13. Air mobility operations, including aerial refueling and air-drop, will remain both challenging and increasingly critical for supporting Air Force and US joint force operations. (Courtesy US Air Force.)

Combined materiel and launch costs for such systems are as low as $100K. They are drawing extensive interest from universities, companies, and others around the world, and numerous such systems have been launched. Efforts are under way to develop low-cost, standardized, on-demand orbital imaging systems for such small satellites. They represent a further aspect of the increasing low-cost access to space that is available to numerous potential adversaries.

Direct-Ascent Antisatellite Capabilities

Offensive counterspace capabilities are also growing, most notably in antisatellite (ASAT) warfare capabilities. In 2007 China became the third known country with a proven ASAT capability when it conducted an unannounced launch of a modified DF-21 intermediate-range ballistic missile, designated the KT-1 space-launch vehicle, to destroy its own defunct Feng Yun-1C meteorology satellite. Essentially any nation with a space launch capability could potentially field a rudimentary ASAT program. The dual use of civilian and military rockets being developed by several countries, including Iran, North Korea, India, and Israel, could lead to rapid growth in the number of players with the technical capability to develop ASAT systems.

Directed-Energy ASAT Technologies

China's demonstrated direct-ascent ASAT capability comes in addition to open-source reports of Chinese and Russian efforts to develop directed-energy ASAT capabilities, primarily in the form of ground-based lasers. Even relatively low-power lasers can temporarily blind or dazzle EO and IR sensors in ISR satellites. At somewhat higher power levels, they can permanently disable these detector arrays or can degrade the effectiveness of solar panels and other components to render a satellite inoperable. At sufficiently high power, they can induce thermal overloads in satellites or even destroy them directly. The rapid time scales on which these effects can be achieved make satellite self-protection extremely challenging. In principle such ground-based laser ASAT systems can be made mobile and widely dispersed, further complicating satellite protection at the terrestrial end.

Co-orbital ASAT Systems

Beyond direct-ascent kinetic-kill ASAT capabilities and ground-based lasers for directed-energy ASAT capability, co-orbital satellites represent a further emerging ASAT concern. Such systems might be relatively small satellites designed to provide an on-demand kinetic-kill capability or might have laser-based or microwave-based DE capabilities to degrade or destroy satellites. Co-orbiting satellites could also provide an adversary counterspace options beyond ASAT capability by, for instance, interfering at relatively close ranges with satellite uplink and downlink transmissions. Such small, maneuvering, co-orbiting satellites also offer many other options for lethal and nonlethal "proximity operations" in support of counterspace efforts.

High-Altitude Nuclear Detonation Effects on Space Systems

Beyond the well-known prompt electromagnetic pulse (EMP) effect of a high-altitude nuclear detonation that affects terrestrial and airborne electronics, a further long-term effect can specifically disable space systems. Such a nuclear detonation would act to populate Earth's Van Allen radiation belts with large numbers of energetic electrons produced from beta decay of fission fragments. These high-energy electrons would remain trapped for years by Earth's magnetic field. Satellites in LEOs or highly elliptical orbits (HEO) would be disabled from effects of the ionizing electrons on critical satellite parts over months or years as they pass through the resulting enhanced radiation belts. This could occur as a side effect of a regional nuclear war or as a deliberate effort by a rogue nuclear adversary seeking to cause massive economic damage to the industrial world.

Figure 14. Increasingly capable RPA systems (*left*) are being used worldwide, while DE systems (*right*) may achieve tracking, targeting, and destruction of such systems. (Courtesy US Navy [*left*] and US Air Force [*right*].)

Orbital Debris

Increasing amounts of orbital debris also pose serious threats to space assets, as demonstrated by the February 2009 collision between a functioning Iridium-33 satellite and a defunct Russian Kosmos-2251 satellite. The collision itself produced nearly 400 new pieces of orbital debris. By comparison the 2007 Chinese kinetic-kill ASAT demonstration produced between 20,000 and 40,000 debris pieces. There are in fact some 1,600 conjunctions predicted every day within the accuracy of current analyses. Just two weeks prior to the Iridium-Kosmos collision, a discarded 4-ton Chinese upper-stage rocket body passed within 48 meters of the 8-ton European Envisat Earth-observation satellite, producing a one-in-70 chance of a collision that by itself would have doubled the entire catalog of human-made objects in orbit.

Cross-Domain Vulnerabilities of Air, Space, and Cyber Systems

Space has long been recognized as an operational domain with capabilities far beyond the original supporting functions that it provided for the air domain. The cyber domain as well has come to be recognized as a domain of its own beyond the supporting functions that it originally provided for the air and space domains. Yet Air Force systems and operations in the air, space, and cyber domains are highly interdependent. Resulting cross-domain effects associated with this interdependence have only recently begun to be understood. They include opportunities for gaining synergies through properly orchestrated combined operations in all three domains to achieve more than the sum of individual operations. However, they also include far-reaching cross-domain threats that arise from these interdependences, many of which are only recently being widely recognized. Potential adversaries will increasingly seek to exploit these cross-domain influences. The challenge is to continue gaining the synergistic cross-domain benefits of these interdependences while minimizing the potential vulnerabilities that they can create.

Cyber Domain Challenges

While the cyber domain was originally envisioned mainly in the context of computer software, hardware, and networks, it is now recognized as encompassing the entire system that couples information flow and decision processes across the air and space domains. It thus comprises traditional wired and fiber-optic computer networks based on electromagnetic (EM) waveguides, but also free-space wireless transmissions of voice, data, and video at radio frequencies, optical wavelengths, and other parts of the EM spectrum, in addition to the hardware and software associated with such systems. Viewed in this way, the enormous cross-domain opportunities and threats presented by the cyber domain become more clearly apparent.

Coupling of Cyber, EW, and ISR Vulnerabilities

Cyber domain opportunities and vulnerabilities apply not only to traditional network warfare and information warfare but also extend to all aspects of EM spectrum warfare. The latter encompasses certain aspects of EW, some of which in turn have close synergies with ISR and other operations in both the air and space domains. This strong coupling of all three Air Force operational domains through the EM spectrum reveals how essential freedom of operation and control of the spectrum will be both for protecting against cross-domain vulnerabilities and enabling the potentially enormous benefits that can come from cross-domain synergies.

Potential adversaries have recognized these cross-domain vulnerabilities and the resultant opportunities that could be presented to them. Because US joint force operations have evolved to become highly dependent on the cyber domain, the strong integration and interdependence of all three Air Force operational domains provide most potential adversaries with an inherently asymmetric advantage. They have found that attacks through cyberspace allow them to create tactical, operational, and strategic effects at low cost and with relative impunity. Most see the cyber domain as a relatively inexpensive venue within which they can potentially offset many of the Air Force's current overwhelming advantages in the air and space domains.

For instance, as China undertakes its force modernization, it does so at a time when cyber threats and opportunities are being understood. It thus has an asymmetric advantage in being able to build a military cyber infrastructure that could potentially be less vulnerable than that which had been put in place much earlier by US joint forces. Developing the ability to make Air Force operations resilient to cyber attacks—as distinct from the current emphasis on protecting adversary access to cyber systems—will become increasingly important.

Cyber Operations in Untrusted Environments

Other countries are increasingly becoming the majority providers of equipment and infrastructure for cyber systems. Particularly in cyber networking technologies, markets have judged foreign-based sources to be comparably capable and less costly than US providers. This situation is causing critical mission traffic to pass over enterprise architectures that involve significant foreign control and that therefore cannot be fully verified with regard to trust/integrity. This includes foreign suppliers of both hardware and software, creating risks to critical missions that must operate in inherently untrusted cyber environments. Although these cyberspace systems can greatly enhance mission effectiveness, a clear understanding of the resulting vulnerabilities they create in key mission areas is typically lost in the rush to gain the benefits of cyber-enabled systems. As Air Force capabilities across the air, space, and cyber domains become increasingly reliant on such systems, a key challenge will be to effectively manage the risks associated with having to operate in environments that are inherently untrusted.

Strategic Implications of S&T Globalization

Further aspects of the strategic context affecting Air Force S&T during this time originate from changes that are occurring due to globalization of science and technology. Over the next two decades, the United States will not dominate science and technology—or the world economy—the way it has over the past 50 years. Already today, the substantial S&T gap that once existed between the United States and much of the rest of the world is closing rapidly. Some say this is the result of US inability to continue educating and advancing a competitive workforce in science, technology, engineering, and mathematics (STEM) fields. Yet regardless of how well the United States addresses that challenge, it is at the other end of the "gap" that most of the change is occurring.

The rest of the world has recognized science and technology as central to national economies that can compete effectively in the knowledge-based global marketplace. They have sent their best and brightest to be educated, often at US universities, and many have returned home to build education systems that, over time, have allowed development of substantial indigenous science and engineering workforces. Some of these countries have also used internal resources to develop the industrial capacity needed to translate science and engineering knowledge into technology-derived products and capabilities. The effect of these investments is that the worldwide S&T gap will continue to close regardless of what the United States may do. The United States and DOD will be competing on an S&T playing field that is substantially different than it was a decade or two ago.

In parallel with this, knowledge and information have become readily available worldwide. Just a decade ago, gaining access to the technical papers needed to understand the results of prior development efforts on any given topic around the world, and thereby identify the most productive avenues for further work, was a significant challenge and a substantial barrier to entry in most technical fields. Today the Internet puts literally thousands of technical papers published every day at nearly anyone's fingertips. The literature search needed to understand the status of any given field and determine the best ways to advance it may be only a few mouse clicks away. The net effect is a rapid and massive global diffusion of science and engineering knowledge that has fundamentally changed how technology advancements made on one side of the world can be translated into militarily relevant systems on the other.

At the same time there has been a growing convergence of the key technologies involved in worldwide consumer products with the technologies needed for enabling many advanced military systems. This trend has been driven primarily by the role that advanced electronics technologies—once largely the domain of high-end military hardware—have come to play in the consumer products market. Today, the technical level of the electronics, processing, and software in many ordinary consumer devices may differ only marginally from those in many high-end military systems, and in some cases exceeds these.

The effect has been to create a global science and engineering workforce having technical skills that can be readily converted from developing consumer products to developing substantial military capabilities. Similarly, enormous investments being made worldwide in legitimate life sciences and biotechnology research are providing the knowledge and workforce that can enable others to develop closely related biological weapons capabilities in a relatively short time. In short, the advancing globalization of S&T is bringing knowledge, technology, and industrial capacity developed for the consumer market within reach of a far wider range of players who can readily transform these into militarily capable systems. Others will find easy access to the resulting technology-derived military systems in international arms markets.

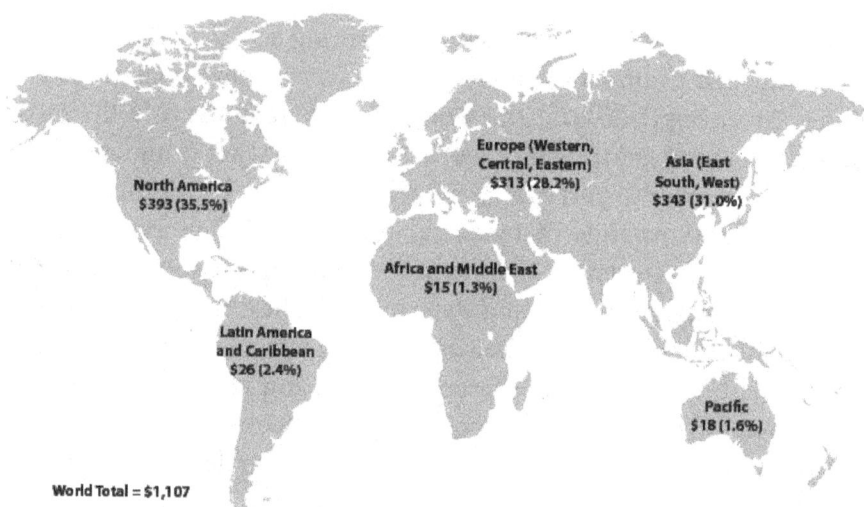

Figure 15. Regional research and development (R&D) expenditures in billions of dollars and percentage of world total, showing North America, Europe, and Asia as roughly equal in R&D investment. (Adapted from National Science Board, *Science and Engineering Indicators 2010* [Arlington, VA: National Science Foundation, 2010].)

A further effect of the globalization of S&T and the resulting rapid pace at which technical advances are being made is that the lifetime of many such advances is becoming remarkably shorter. In part due to the sheer number of scientists and engineers now working worldwide in technical fields, the pace of both technical innovation and obsolescence has increased sharply. In the consumer products market, it is the innovation cycle that receives the most attention, but in the military systems market it may be the obsolescence cycle that is more important. As a result, the lifetime of at least some technology-derived military systems may not be nearly as long as it has been in past decades. This can be expected to impact defense acquisition choices as well as the S&T investment decisions needed to support these.

Federal Budget Implications for Air Force S&T Strategy

A final essential aspect of the strategic context that will guide Air Force S&T during 2010–30 is the influence that federal budget pressures will have. Combined impacts of the federal debt, the aging population, and federal entitlement promises made through Social Security, Medicare, and Medicaid have begun what will become over the next several decades an unprecedented drain on the federal budget that will increase rapidly in coming years. The first of 77 million retiring baby boomers born between 1946 and 1964 became eligible for Social Security benefits in January 2008. Their numbers will grow at a rate of four million per year through 2026, and they will continue to draw entitlement benefits through 2050.

The Congressional Budget Office (CBO) has calculated that the cost of these benefits will grow from 8.4 percent of the gross domestic product (GDP) today to 14.5 percent by 2030, and 18.6 percent by 2050. By comparison, the entire federal budget today is just 20 percent of the GDP. At current rates, the CBO has estimated that by 2049 these entitlements would consume every other federal program except interest on the federal debt. Even with potential changes in entitlement benefits, the looming federal budget pressures will be immense and will necessarily impact the range of S&T and other Air Force efforts that can be pursued.

Chapter 3

Enduring Realities for the Air Force
2010–30

This chapter notes several enduring factors that will continue to influence Air Force S&T needs and the environment in which Air Force S&T will be conducted over the next two decades. Beyond elements of the strategic environment in the previous chapter, additional key factors represent enduring needs of the Air Force in 2010–30 and form essential background for an S&T vision to guide investment choices over this period.

Ensuring Interoperability with Legacy Systems

In 2030 many manned, RPA, and other systems that the Air Force operates today will continue to be needed and remain in use. Ensuring interoperability with these systems, especially as the technologies in newer systems advance rapidly, will become an increasing challenge. Technologies that can be cost-effectively integrated in these systems to upgrade their capabilities, or that otherwise enable these legacy systems to interoperate effectively with newer systems, will become even more important than they are today.

Sustainment Costs for Legacy Systems

Sustainment is essential to the Air Force mission. As legacy air platforms and other systems continue to be used throughout this period and as new platforms and systems are introduced during this time, technologies to support improved sustainability or to reduce costs associated with sustainment will continue to be essential.

Importance of Low-Observable Systems

LO technologies and the systems that employ them for long-range penetrating and persistent strike are among the most distinguishing elements of the Air Force. While advanced IADSs may over time create an increasingly challenging environment for these critical systems, LO systems will remain essential for the ability they give to penetrate defended airspace, for the sensitivities they demand in the air-defense systems of potential adversaries, and for the potential secondary benefits that this can create for other technology-based capabilities. Technologies to extend affordable LO capabilities will remain essential.

Energy Costs and Availability

Reducing Air Force energy costs, especially the costs associated with aviation fuel usage, and increasing the assured availability of energy, including fuels and base energy, will continue to be growing challenges during 2010–30. As with other Air Force costs, fuel and energy outlays will reduce the availability of resources that could otherwise be used to provide increased capability dominance. The need for advances in a broad range of technology areas to reduce fuel use—including propulsion, structures, aerodynamics, and controls, as well as technologies to allow increasing use of alternative fuels—will remain important, as will technologies to enable air base operations with increasingly greater independence from the electricity grid.

Growing Role of the Cyber Domain

During the past decade, no other technology domain has grown in importance as rapidly as the cyber domain. Understanding of potential vulnerabilities that cyber systems create has grown from defending computer networks to include network-enabled RF systems and other parts of the EM domain, as well as hardware and software supply-chain vulnerabilities. This domain has become centrally important for maintaining Air Force mission assurance and for enabling entirely new mission capabilities. While the defensive aspects of cyber warfare may improve somewhat as systems developed in the "precyber" age are retired, cyber warfare at various levels will continue to grow rapidly in importance in the 2010–30 environment.

Need for "Soft Power" Capabilities

Air Force missions have always included many types of other-than-conflict operations, including emergency response, humanitarian relief, peacekeeping missions, and contingency operations as well as joint exercises and other cooperative activities with international partners. The role of these missions has grown in recent decades and will likely continue to expand over the next two decades. There will be an enduring need for technologies that can enable a wide range of Air Force capabilities that support these types of missions, including better multilevel security solutions to facilitate sharing of information, systems, and training with international partners.

Manpower Costs

Manpower costs associated with pay and benefits are consuming a significantly growing fraction of the total Air Force budget, pressuring resources for all other functions that support Air Force missions. Technologies that can enable the same or greater mission effectiveness with reduced manpower requirements can free up resources to permit greater overall Air Force capability. For example, these may include technologies for increasingly autonomous systems and methods to increase trust in them; for augmentation of human performance; for improved processing, exploitation, and dissemination (PED) of ISR data; and for reducing maintenance and sustainment costs. The need to focus S&T efforts on these and other technologies to reduce manpower costs will continue to become increasingly important.

Budget Constraints

Air Force budgets throughout this period will likely remain constrained. As noted in chapter 2, combined impacts of the federal debt, the aging population, and federal entitlement promises made through Social Security, Medicare, and Medicaid have begun an unprecedented drain on the federal budget that by 2030 will consume nearly three-quarters of the entire budget. Even if reductions in entitlement promises are enacted and federal taxes as a percentage of the GDP are raised substantially over today's rates, the effect of these factors will be to place increasingly severe limits on Air Force budgets over this period.

Effective National S&T Partnerships

The Air Force has consistently maintained working partnerships that include coordination of science and technology programs as well as collaborations on joint S&T efforts with its sister services, the Defense Advanced Research Projects Agency (DARPA), National Aeronautics and Space Administration (NASA), Intelligence Advanced Research Projects Activity (IARPA), and other organizations. The importance of such effective partnerships will continue—and is likely to grow—as budget pressures necessitate even greater efficiencies and as the scale and scope of many ambitious efforts exceed the reach of any one organization alone. These benefits of such effective partnerships extend as well to FFRDCs and to industry involved in S&T-related efforts.

Continued International S&T Cooperation

International collaboration in S&T has grown significantly over the past decade, providing access to capabilities that complement those domestically available. Building and expanding international S&T-related partnerships have been achieved, for example, through efforts such as the Engineer and Scientist Exchange and the Windows on Science Programs; the Air Force Office of Scientific Research's (AFOSR) European (EOARD), Asian (AOARD), and Southern (SOARD) Offices of Aerospace Research and Development; and the deputy under secretary of the Air Force for international affairs (SAF/IA) air armaments cooperation program. In addition to leveraging the limited resources available for advancing the US technological base, these international cooperations help build partner capacity and can help in efforts to avoid capability surprise. Such international S&T cooperation programs will play an increasingly important role in the 2010–30 environment.

Science, Technology, Engineering, and Mathematics Workforce

Air Force capabilities are dependent on a workforce that has the needed STEM backgrounds to provide the S&T advances on which these capabilities will be based. Development and maintenance of such a workforce will be essential for achieving the S&T vision laid out in *Technology Horizons*. This applies to the workforce needed within the Air Force Research Laboratory (AFRL) as well as within the broader industrial base. The Air Force already makes important contributions to successful STEM programs focused at the K-12 and college/university levels through open-house efforts in AFRL and via direct mentoring relationships. Support of the Air Force's "Bright Horizons" plan for a coordinated STEM workforce development effort will need to be sustained to provide the workforce needed over the next decade and beyond.

Chapter 4

Overarching Themes for
Air Force S&T 2010–30

*This chapter identifies a number of broad themes that will
guide Air Force S&T in ways that can accelerate the delivery
of technology-derived capabilities able to respond to the strategic environment, budget environment, and technology environment during this period.*

The strategic and budget environments outlined in earlier chapters,
together with the state to which key technologies can be brought during the next decade to meet the demands of those environments, lead
to a set of broad technology themes that will be of overarching importance during this time. These themes provide guidance and direction
for identifying S&T efforts that will be among the most essential in
terms of their impact on Air Force capabilities.

These overarching themes are presented in figure 16. Each row represents a significant shift in research emphasis within a given area,
namely a decrease in emphasis on one area toward an increased emphasis on a different focus area. Both the reduced emphases in the left
column and the increased emphases in the right column combine to
ensure an S&T portfolio that is well suited to enabling key capabilities
the Air Force will need during 2010–30.

The trends in figure 16 are fundamental shifts in research focus areas
to guide Air Force S&T in directions that will provide it with the technologies needed to enable capabilities that are matched to the environment over the next decade and beyond. The identified shifts in research
balance should not be interpreted as a call for ending all research in the
technical areas identified on the left, since this would create a substantially unbalanced S&T portfolio. Instead, greater attention must be
given to those areas on the right.

Beyond the impact on S&T efforts, the shifting emphases in figure 16
point to a need for systems engineers in product centers to become fa-
miliar with the resulting technologies and to develop approaches and

From decreased emphasis on this ...	To increased emphasis on this ...
1. From ... Platforms	To ... Capabilities
2. From ... Manned	To ... Remotely piloted
3. From ... Fixed	To ... Agile
4. From ... Control	To ... Autonomy
5. From ... Integrated	To ... Fractionated
6. From ... Preplanned	To ... Composable
7. From ... Single domain	To ... Cross domain
8. From ... Permissive	To ... Contested
9. From ... Sensor	To ... Information
10. From ... Strike	To ... Dissuasion
11. From ... Cyber defense	To ... Cyber resilience
12. From ... Long system life	To ... Faster refresh

Figure 16. Overarching themes that will guide Air Force S&T during 2010–30 to accelerate delivery of technologies appropriate to the changing spectrum of threats. Emphasis should decrease—but not end—in research supporting the areas on the left to accommodate increased emphasis in those on the right. These trends represent shifts in research emphases, not necessarily in near-term acquisition priorities. (Courtesy Office of the Chief Scientist of the Air Force.)

tools for integrating them in Air Force systems. Design protocols, subsystem interactions, and other aspects of system development all will be impacted by these changing characteristics of future Air Force systems.

From Platforms to Capabilities

Since the end of the Cold War and its single major nation-state threat scenario, emphasis has been increasingly shifting from "platforms" exquisitely designed to overcome a single threat toward "capabilities" able to meet a far wider range of threats. For instance, potential adversary A2/AD capabilities and long-range AAMs are increasing emphasis on standoff strike and beyond-visual-range engagements that rely less on platform-specific functions, such as thrust vectoring for combat maneuverability, and more on capabilities derived from the platform contents, such as ISR and EW functions. This trend is likely to continue, and it suggests that core technologies associated with platforms, such as aerodynamics and flight control, will continue to be needed but that greater emphasis will shift toward other technology areas that deal more directly with capability-providing functions.

> *Although platforms will remain significant drivers of capability, the challenges the Air Force faces require S&T to place increasing emphasis on capability-based efforts over platform-centric efforts.*

From Manned to Remotely Piloted

The growth in Air Force and other DOD use of remotely piloted vehicles has been rapid, as their utility for operations far beyond surveillance has come to be understood. Such systems are also being used for an increasingly wider range of nonairborne applications. Although some types of emerging systems will still benefit from having one or more crew members on board, for many applications the weight, volume, and endurance limitations associated with such crew, as well as the cockpit environmental control systems needed to support crew operations, extract an unacceptable performance penalty. Manned systems also present risks to crew that can be avoided in remotely piloted systems.

For these reasons, an increasing shift to remotely piloted systems is inevitable over the next decade, and technologies will be needed to permit the potentially greater functionality of such systems to be exploited for entirely new missions. Such systems are likely to involve varying degrees of autonomy, require endurance far beyond what is possible today, and include large high-altitude airships with massive ISR capabilities, all requiring new supporting technologies.

> *Technologies to enable an increasingly wider array of remotely piloted systems, many of which will bear little resemblance to today's systems, will become increasingly important over the next decade and beyond.*

From Fixed to Agile

The Air Force is likely to face a broad spectrum of conflict and non-conflict operations in this period. This will demand a shift from systems designed for fixed purposes or limited missions to ones that are inherently agile in their ability to be readily and usefully repurposed across a range of scenarios. In many cases, this is likely to entail less than optimal performance for a primary purpose in exchange for acceptable utility and performance across these broader uses.

This extends also to the range of operating environments in which such systems will be used. Frequency agility, for example, is virtually certain to become an essential attribute as further portions of federal or shared spectrum worldwide are transferred to primary commercial use. Similarly, frequency agility may enable approaches such as pulse-to-pulse waveform diversity that can assist in offsetting the threats posed by DRFM and other electronic countermeasures. A further example of agility is represented by systems in which sensitive functional components can be more easily removed to allow them to be readily shared with international partners or to assist in overcoming multilevel security issues.

Technologies and enabling design principles will be needed to permit such increasingly agile systems to retain the greatest functionality in their primary role(s) while allowing the greatest range of intended secondary roles and, to the greatest extent possible, adequate functionality even under unanticipatable scenarios. To a degree far exceeding what is possible today, Air Force systems will need to become increasingly multirole-capable. The benefits of agility reach well beyond the scenarios identified above and will allow many systems to swing from high-end general-purpose applications to lower-end irregular warfare applications.

> For reasons both of cost savings and flexibility of use, it will become increasingly necessary to develop technologies specifically intended to enable substantially greater agility in systems.

From Control to Autonomy

Airborne remotely piloted systems and many nonairborne systems have significant autonomous capability. However, increasing levels of flexible autonomy will be needed for a wider range of Air Force functions during 2010–30 and beyond. These will include fully unattended systems with reliable levels of autonomous functionality far greater than is possible today, as well as systems that can reliably make wide-ranging autonomous decisions at cyber speeds to allow reactions in time-critical roles far exceeding what humans can possibly achieve.

Key attributes of such autonomy include the ability for complex decision making, including autonomous mission planning, and the ability to self-adapt as the environment in which the system is operating changes. Flexible autonomy refers to systems in which a user can specify the degree of autonomous control that the system is allowed to take on and in which this degree of autonomy can be varied from essentially none to near or complete autonomy.

Limited autonomy is already being widely used in Air Force systems, but their autonomous functionality is far below what could be achieved over the next decade. From a purely technical perspective, methods already exist to extend autonomy to far higher levels than are being used today. Even this level of autonomy will be enabled only when automated methods for verification and validation (V&V) of complex, adaptive, highly nonlinear systems are developed. In the near term to midterm, developing methods for establishing certifiable trust in autonomous systems is the single greatest technical barrier that must be overcome to obtain the capability advantages that are achievable by increasing use of autonomous systems.

Although humans will retain control over strike decisions for the foreseeable future, far greater levels of autonomy will become possible by advanced technologies. These, in turn, can be confidently exploited as appropriate V&V methods are developed along with technical standards to allow their use in certifying such highly autonomous systems. This area is at the interface between complex nonlinear systems theory and the massive computational analytics enabled by the availability of near-infinite processing and storage capacity at near-zero unit cost.

It is thus reasonable to expect that with increasing S&T emphasis in this area, technologies will be developed over the next decade that can enable reliable V&V and certification methods to provide trust in even highly adaptable autonomous systems. These, in turn, can open up entirely new avenues for reducing manpower needs and entirely new missions and capabilities for Air Force systems. Such trusted highly autonomous systems will be essential for dominance in the cyber domain, where operation at the speeds that autonomy offers will be combined with levels of autonomous control far beyond what is possible today.

Note that potential adversaries may be willing to field highly autonomous systems without any demand for prior certifiable V&V. In so doing they may gain potential capability advantages that we deny ourselves by requiring a high level of V&V. One of the main conclusions of this study is that such highly autonomous systems can generally provide enormous capabilities.

> *Emphasis on developing methods to achieve V&V of complex, highly adaptive, autonomous systems will be essential to enabling the capabilities they can provide.*

From Integrated to Fractionated

To date, in most system architectures the various functions needed for the system to operate have been physically integrated via subsystems in the overall system. Thus, for example, most aircraft integrate communications, EW, ISR, strike, and other mission functions together into a single platform. The resulting systems become relatively large and heavy, with associated performance penalties that translate into limits on range and other factors and with relatively high unit production costs and operating costs. Moreover, in such system architectures the loss of any one subsystem can lead to mission failure or even loss of the entire system. From a mission survivability perspective, such architectures demonstrate poor survivability.

Yet, it is becoming increasingly possible to fractionate systems into a relatively small set of functional subsystems, each operating spatially separated from the others but maintaining local communication with them to allow overall system functionality to be preserved. The system is, in effect, separated into fractional elements and physically dispersed. If the fractionation is done correctly, then the short-range communication bandwidth between elements can be made low enough that burst mode transmission, frequency agility, laser links, and other methods can be used to achieve low probability of detection and maintain jam-resistant local link integrity.

Fractionation provides a dispersed architecture that, even by itself, can create cost and risk advantages over traditional integrated systems. It is distinct from modularity in that the latter is designed to more readily allow subsystem replacement but still aggregates subsystems into an integrated architecture. Most importantly, when a fractionated architecture is augmented by even low levels of redundancy among the dispersed elements, survivability can increase dramatically. Combining fractionation with redundancy in a system-level architecture produces survivability that far exceeds what is possible by traditional redundancy alone.

For a system architecture with m different types of functional elements and n copies of each, the number N of shots each having individual probability-of-kill P_k that are needed to cause mission-loss probability P_L can be readily shown to be given by

$$N = \frac{\log\left(1 - \left[1 - \left(1 - P_L\right)^{1/m}\right]^{1/n}\right)^{mn}}{\log\left(1 - P_k\right)} .$$

Any fractionated architecture corresponds to particular m and n values, and the traditional integrated system architecture corresponds to the trivial case where $m = n = 1$. Using the above equation it can be readily shown that, for reasonable P_k values, the number of shots N needed to cause any mission-loss probability P_L increases rapidly with the degree of fractionation and redundancy. The combination of fractionation and redundancy produces a multiplicative benefit beyond what is possible in a traditional architecture.

Such highly survivable fractionated system architectures are only recently becoming possible through technologies that enable jam-resistant secure local communications and collaborative control to permit the spatially dispersed functional elements to operate as a single coherent system. Fractionated architectures are a key to the development of low-cost autonomous systems that can survive in A2/AD environments. Such architectures and the key technologies that enable them have been relatively unexplored to date.

> Fractionated systems represent a relatively new technology-enabled architecture for achieving increased system-level survivability with low-cost autonomous system elements.

From Preplanned to Composable

Composability refers to the ability to rapidly assemble an as-needed mission-specific capability from an available set of standard and relatively low-cost elements. It raises the concept of agility from the level of a single system to the entire set of systems needed for mission-level capability. By composing a mission-level capability from a standard set of common elements, a broader range of missions can be effectively addressed at lower cost than is possible with a more traditional aggregated system designed for a specialized or narrow range of missions.

For example, the same fractionated mission elements noted above could be used to compose a relatively simple EO surveillance mission in uncontested airspace using just one element having imaging capability. The same set of elements could be used to compose a hunter-killer mission package by adding the strike variant. At the other extreme, for heavily defended airspace a large number of EO, EW, strike, and other variants could be composed into a package suitable for that mission. The key point is that for a wide range of mission types, it may be possible to compose a "good enough" mission package from the appropriate mix of a set of common-core, flexibly autonomous elements that differ primarily in their function-specific payloads.

The underlying concept of composability can be applied broadly. In cyberspace, for example, by storing elements from a common set of functional cyber-mission components in a network in advance of a cyber attack, these elements can be very rapidly composed into a good-enough mission package to respond to the attack. The point is that composing a mission package from an already available set of common elements can allow for a faster response at lower cost that is "good enough" to deal with a much wider range of missions.

In the cyber domain, the fundamental inability to anticipate every possible attack and preplan an ideally tailored response to each is readily apparent. However, the same idea of a mission package composed as needed from preexisting low-cost elements is central to the previous example as well. Similarly, plug-and-play modularity to enable rapid custom satellite composition is a further embodiment of composability, and the approach can even be extended to as-needed composition of good-enough small satellite constellations for operationally responsive space missions.

> *Gaining the cost advantages and mission flexibility that are available through composability will require developing technologies to enable composable mission capabilities in all domains.*

From Single Domain to Cross Domain

Air Force systems are broadly thought of as belonging in the air, space, and cyber domains, yet it is being increasingly recognized that key vulnerabilities can arise in these systems as a result of cross-domain influences and interdependences. It is possible to exploit cross-domain effects in adversary systems as well or even to derive beneficial effects from cross-domain influences. Here, cross-domain effects are defined to include a technology in one domain that can produce unexpected beneficial or detrimental effects in another domain, or a technology in one domain that requires supporting functions in another domain and as a result creates interdependences between two or more domains, or even a technology that falls between classical domains but has implications in one or more of them.

In general, the awareness of these cross-domain effects is only beginning to be understood, in part as a result of recent rapid growth in vulnerabilities of cyber and space systems. Cross-domain effects apply, however, in all three domains. Over the next decade, efforts to develop new technologies will need to increasingly take into account possible unintended cross-domain effects. At the same time, technologies to minimize existing cross-domain vulnerabilities and to exploit similar vulnerabilities in adversary systems will be increasingly needed.

> *Emphasis will increase on technologies to understand, anticipate, and, where appropriate, avoid cross-domain effects in Air Force systems and to exploit them in adversary systems.*

From Permissive to Contested

With only a few exceptions, Air Force operations over the past two decades have been conducted under largely permissive conditions in relatively poorly defended airspace. This risks the development of an underappreciation for the impact that possible future operations in A2/AD environments could have on some missions and functions, including those in the cyber domain. Greater emphasis should be placed on technologies that can operate effectively in moderate or heavily contested domains.

This will require developing technologies specifically designed to enable resilience of key Air Force mission functions in contested domains. In developing technology solutions for any specific problem, it will be essential to ensure that these have the needed resilience to operate effectively even in significantly contested domains. Examples include the need for EW technologies that can be effective in situations where control of the EM domain is not assured. Similarly, technologies will be needed that can not only permit RPAs to communicate and operate under threats in defended airspace but also permit other weapon systems to achieve greater range and deep penetration ability. In the increasingly contested space domain, technologies will be needed to ensure resilience of space assets and space operations to various forms of threats.

> *To meet the needs of Air Force operations during this period, it will be essential for S&T emphasis to increase on technologies that can remain effective even in contested domains.*

From Sensor to Information

As a result of rapid advances made in electronics and in imaging technologies in particular, there has been an enormous surge in the amount of sensor data available across many portions of the EM spectrum. Current EO-IR sensors, for example, generate huge amounts of raw imaging data, and similar increases in raw data streams are occurring from other sensing modalities, including hyperspectral imaging, light detection and ranging (LIDAR) imaging, electronic signals intelligence, and other RF-based ISR sensing modes. The volume of sensor data from current-generation sensors that must be processed, exploited, and disseminated has become overwhelming, as manpower requirements to deal with these data have placed an enormous burden on the Air Force.

It appears clear that the focus on developing ever more capable sensor technologies of the present type is far outstripping the ability to deal efficiently with the resulting raw data. Such current-generation sensors can be characterized as "dumb" insofar as they are designed solely to generate raw sensed data, rather than to generate interpreted information from these raw data. Over the next decade, development focused solely on higher-output sensor technologies will need to be deemphasized to accommodate increased emphasis on developing technologies for "intelligent" sensors. These will, for example, provide cueing-level processing of raw data on the sensor itself and then transmit only those data segments containing cues to the ground.

Technologies that make use of continuing advances in miniaturization of electronics and the availability of massive processing and storage in extremely small volumes can, in principle, allow limited processing of raw sensor data to be done behind the sensor itself. This offers potential to transform sensor output from raw data to real information. Processing done on the sensor need not be as detailed as that currently done on the ground and might be used only to cue burst transmissions of raw data to the ground for subsequent detailed processing. Resulting substantial decreases in the amount of raw data being transmitted would alleviate bandwidth requirements as well as ground-based PED manpower requirements.

Beyond development of such intelligent sensors that provide information rather than data, other technologies may assist with increasingly autonomous data processing on the ground to reduce current PED manpower requirements. As with intelligent on-sensor processing, such autonomous off-sensor processing may not need to be as detailed as current manual processing since it might only be used to cue raw data segments for subsequent manual processing. The combined impacts of intelligent sensors to reduce raw data streams and autonomous ground processing to assist in current PED functions can potentially be enormous.

In general, Air Force S&T emphasis in the next decade will need to shift focus from advancing current dumb sensor technologies to advancing the ability to extract useful information from the sensor data stream. The utility of ever-greater situational awareness created by expanded ISR systems will make this an exceptionally high priority for the S&T community.

> A significant shift in Air Force S&T emphasis is needed toward developing technologies that can increasingly assist in converting raw sensor data into useful information.

From Strike to Dissuasion/Deterrence

Dissuasion and deterrence have played a role in the Air Force mission since its inception as a distinct service. However, the nature of the broader deterrence mission that applies in today's environment, and that is likely to become increasingly important during 2010–30, is substantially different. Future strategy will include nuclear and conventional components and must address state and nonstate actors as well as other entities that are not dissuaded or deterred by traditional means based solely on threats of retaliation or punishment. The corresponding tools needed to deal effectively with that environment are also different. In addition to technologies needed for active conflict operations, there will be a need during this period to enable appropriate dissuasion and deterrence tools for that environment.

As noted in chapter 2, deterrence in the context of 2010–30 involves not only effectively deterring nations that are pursuing nuclear-weapons capabilities, such as North Korea and Iran, but also deterrence in the conventional warfare sense, as well as in the cyber domain and even deterrence of chemical and biological warfare arenas. The broad spectrum of the Air Force role in these missions and the wide range of actors involved create a need for significantly different and more effective tools that can support these deterrence roles.

In many cases, to be effective these tools will require nonkinetic means to exert influence on potential adversaries. The cyber domain offers significant capabilities for exerting a deterrent influence on some such actors. Deterrence in the context of what is often termed *irregular warfare* and *hybrid warfare* brings special difficulties that may require new approaches. Actors other than nation-states may be less susceptible to such influence, and other tools will be needed to deal with them.

In general, S&T will need to play an expanded role in developing technologies that can support the Air Force role in providing effective dissuasion and deterrence capabilities across the spectrum of conflict and nonconflict scenarios. These efforts may benefit from coordination with the other services, with DARPA and IARPA, and with our international partners.

> *Over the next decade, technologies will be needed that can enable effective deterrence capabilities across a far broader spectrum of threats than has historically been the case.*

From Cyber Defense to Cyber Resilience

Since the earliest awareness of the vulnerabilities of cyber systems and the far-reaching implications of them, efforts to control these have been based largely on cyber defense. This refers to the many traditional approaches intended to prevent adversaries from entering computer networks and other cyber-accessible systems. Yet despite extensive efforts to block adversaries from entering these systems, they can and routinely do access them. Today the efforts that must be expended to protect cyber systems in this manner far exceed those that adversaries expend to penetrate them. There is essentially no evidence to suggest that this will change significantly as long as the focus remains largely on cyber defense.

Entirely new approaches to assuring continued mission effectiveness of cyber systems will be needed, allowing us to better fight through cyber attacks. A key step is to begin a fundamental philosophical shift from emphasizing cyber protection to emphasizing mission assurance. This subtle but important change opens vastly different ways to approach the cyber challenge. For example, mission assurance can be dramatically increased by technologies that can inherently make cyber systems more resilient to the presence of adversaries within them. S&T efforts to develop technologies for increased cyber resilience will be a key theme in the coming decade and one of the cornerstones for achieving mission assurance.

For instance, as described in chapter 5, massive virtualization coupled with rapid network recomposability can make cyber systems inherently far less vulnerable to direct network attacks. This technology represents a fundamental shift in the approach to network defense, replacing current efforts to keep adversaries out with a new paradigm that makes it nearly useless for an adversary to gain entry. The resulting network thereby becomes inherently resilient to attack.

Such inherent cyber resilience via massive virtualization and rapid network recomposability can also cause cyber adversaries to leave behind far greater forensic evidence for attribution, regardless of whether they entered via the network layer, backdoor implants in hardware, or the application layer. The technology can provide inherent resilience in the sense that its benefits do not require any explicit awareness of the adversary's presence in the system or explicit action by the defender to gain the benefit.

> The fundamental shift in focus from cyber defense to cyber resilience represents one of the most far-reaching themes for Air Force S&T in the coming decade.

From Long System Life to Faster Refresh

It is a reality that many of the air platforms and other systems that the Air Force employs will continue to be needed and will remain in use long after they are first introduced into operation. Yet such long-life systems often make integrating new technology-derived updates into them inherently difficult and costly. This is particularly true when systems are not developed with architectures specifically designed to accommodate such future upgrades. At the same time, technologies continue to advance at an increasingly faster pace, and worldwide access to the resulting military systems developed from them is also increasing rapidly.

The result is that the global technology refresh rate is now far faster than just a decade or two ago. Air Force capabilities based on systems designed for long-life use will remain appropriate for some applications. However, adversary capabilities obtained with less exquisite systems may have increasing access to faster technology refresh cycles. For many types of systems, it will become increasingly difficult for the Air Force to maintain sufficient technology advantage without itself having access to substantially faster technology refresh rates.

Technologies need to be developed to allow subsystems, if not entire systems, to be inherently expendable by design in the sense that they are designed and integrated with the specific intent to replace them with newer technology after far shorter usage than is the case today. In part, this requires new technologies that can enable key system functions to be implemented at much lower cost. It also requires advances in technologies for "open architectures" that are secure but can enable ready integration of successive generations of newer subsystems.

Development of technologies to support such innately expendable systems also supports other overarching themes identified in *Technology Horizons*, including agility and composability, as well as the shift toward fractionated system architectures. All of these are essential elements for maximizing US joint capabilities while supporting the need for the Air Force and DOD to move from exquisite systems to more flexible, agile, and broadly effective capabilities that can meet the needs of a wider and increasingly less predictable set of future operational scenarios.

> New approaches that can enable far faster technology refresh rates in Air Force systems and subsystems are key to achieving the far greater flexibility that will be needed to respond to the range of future threats.

Chapter 5

Technology-Enabled Capabilities for the Air Force 2010–30

Potential technology-enabled capability areas are identified that are aligned with key needs of the strategic environment, enduring realities, and overarching themes that the Air Force faces and that can be achieved with realistically attainable technology advancements over the next 10 years.

The notional process used in *Technology Horizons* for identifying "potential capability areas" (PCA) is shown in figure 17. For each of the air, space, and cyber domains, it identifies key functions associated with the domain and uses these to envision PCAs that could provide these domain functions. For each candidate PCA, it considers the technology areas that would be needed to enable that capability and the state to which each technology area would need to be brought in order to provide the potential capability being envisioned.

If a technology area could not be credibly advanced from its current 2010 state to the needed state by the 2020 target in the 10+10 Technology-to-Capability process (see chap. 1, fig. 2), then an alternate technology for enabling that PCA had to be identified, or the capability was rejected. Similarly, if an adversary using the same state of technologies could readily counter a potential capability, or if a capability would create unacceptable vulnerabilities—either by itself or through cross-domain influences it would produce—then it was also discarded. In so doing, *Technology Horizons* envisioned PCAs that would be of value in the 2030 strategic environment and that could be credibly achieved if the needed S&T investments are made.

Note this does not mean that the technologies needed to develop these capabilities are already sufficiently mature or that the Air Force should necessarily acquire these capabilities. Instead, these notional capabilities are identified because they are well matched to needs of the Air Force within the time horizon of the study. This set of PCAs is then used as a mechanism for identifying the KTAs that would be needed to

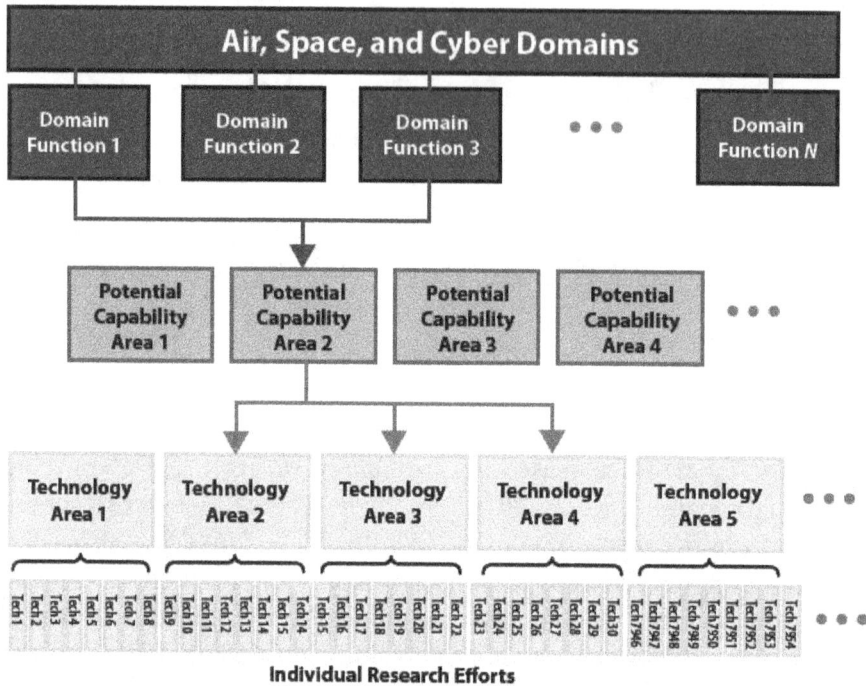

Figure 17. Notional process used in *Technology Horizons* for identifying PCAs and KTAs for the air, space, and cyber domains. (Courtesy of the Office of the Chief Scientist of the Air Force.)

enable them. In chapter 6, mapping the KTAs across the PCAs allows identification of the most valuable technology areas that Air Force S&T should be pursuing now.

This process allowed *Technology Horizons* to maintain its focus on assessing the most valuable technology areas for the Air Force, without specifying what future capabilities the Air Force should be acquiring. Note that the PCAs used in this process should not be interpreted as an alternative to "flagship capability concepts" or "focused long-term challenges" that the AFRL may use as constructs for communicating, planning, and managing its S&T efforts. Instead, the PCAs are a complementary means used here for understanding the relative impacts of various KTAs. They allow identifying the KTAs that have the greatest crosscutting value in enabling capability areas that are aligned with the strategic context and enduring realities of 2010–30.

Alignment of Capability Areas with
Air Force Core Functions

The process described above is predicated on the set of PCAs being sufficiently broad to address most of the key functions the Air Force needs to perform its mission. This ensures that the KTAs determined by this process are the most valuable for Air Force S&T to pursue and can support a sufficient range of the functions needed to meet likely Air Force needs. To assess if this is the case, researchers mapped the final set of PCAs into the Air Force service core functions (SCF) to determine the range of these that the underlying KTAs will support.

Air Force Service Core Functions

The Air Force has defined the following set of 12 SCFs that encompass the combat and support functions it needs to accomplish its range of missions.

- SCF1:   Nuclear Deterrence Operations
- SCF2:   Air Superiority
- SCF3:   Space Superiority
- SCF4:   Cyberspace Superiority
- SCF5:   Global Precision Strike
- SCF6:   Rapid Global Mobility
- SCF7:   Special Operations
- SCF8:   Global Integrated ISR
- SCF9:   Command and Control
- SCF10: Personnel Recovery
- SCF11: Building Partnerships
- SCF12: Agile Combat Support

The final set of 30 PCAs used in *Technology Horizons* is listed in the next section. Figure 18 maps these PCAs across the above SCFs to assess the range of these functions that each area supports.

> *Mapping the PCAs across the SCFs ensures that* Technology Horizons *considered a sufficiently wide range of technologies to support the full set of Air Force missions.*

Technology-Enabled Potential Capability Areas

A set of 30 PCAs was identified as being credibly achievable within the time horizon this study addresses and as meeting key needs defined by the strategic environment, enduring realities, and overarching themes that apply to 2010–30. Each of these PCAs is listed below, and all are summarized in a following section.

- PCA1:   Inherently Intrusion-Resilient Cyber Systems
- PCA2:   Automated Cyber Vulnerability Assessments and Reactions
- PCA3:   Decision-Quality Prediction of Behavior
- PCA4:   Augmentation of Human Performance
- PCA5:   Advanced Constructive Discovery and Training Environments
- PCA6:   Trusted, Adaptive, Flexibly Autonomous Systems
- PCA7:   Frequency-Agile Spectrum Utilization
- PCA8:   Dominant Spectrum Warfare Operations
- PCA9:   Precision Navigation/Timing in GPS-Denied Environments
- PCA10: Next-Generation High-Bandwidth Secure Communications
- PCA11: Persistent Near-Space Communications Relays
- PCA12: Processing-Enabled Intelligent ISR Sensors
- PCA13: High-Altitude, Long-Endurance (HALE) ISR Airships
- PCA14: Prompt Theater-Range ISR/Strike Systems
- PCA15: Fractionated, Survivable, Remotely Piloted Systems
- PCA16: Direct Forward Air Delivery and Resupply
- PCA17: Energy-Efficient, Partially Buoyant Cargo Airlifters
- PCA18: Fuel-Efficient Hybrid Wing-Body Aircraft
- PCA19: Next-Generation High-Efficiency Turbine Engines
- PCA20: Embedded Diagnostic/Prognostic Subsystems
- PCA21: Penetrating, Persistent Long-Range Strike
- PCA22: High-Speed Penetrating Cruise Missile
- PCA23: Hyperprecision Low-Collateral-Damage Munitions
- PCA24: DE for Tactical Strike/Defense
- PCA25: Enhanced Underground Strike with Conventional Munitions
- PCA26: Reusable Air-Breathing Access-to-Space Launch

- PCA27: Rapidly Composable Small Satellites
- PCA28: Fractionated/Distributed Space Systems
- PCA29: Persistent Space Situational Awareness (SSA)
- PCA30: Improved Orbital Conjunction Prediction

A much larger set of PCAs was considered; however, this list gives the 30 PCAs that were assessed as being most valuable for meeting Air Force needs. The list is not prioritized in any way since it was not the objective of the study to define capabilities the Air Force should acquire but to prioritize the most essential technology areas that are needed to enable capabilities such as these. These are PCAs in the sense that most represent systems or functions that enable what would be regarded as future capabilities. These capability areas also serve as example embodiments of the broader overarching themes identified in chapter 4.

Alignment of Potential Capability Areas with Service Core Functions

Figure 18 shows the mapping of each of the above PCAs into each of the 12 Air Force SCFs previously listed. In each case, a check mark indicates that the PCA could significantly assist in the performance of at least one or more aspects of the corresponding SCFs. Where there is no check mark, there may still be connections with that capability area, but these do not rise to as comparably meaningful a level as those that are marked in the figure.

Brief Descriptions of Technology-Enabled Capabilities

Each PCA is briefly described below at a level sufficient to understand the basic capabilities being proposed, the technologies that can enable them, and the benefits they would provide for the Air Force. For each of these PCAs, there is technical support for the claim that the underlying technologies that can be brought to the needed level of readiness by the 2020 technology horizon date in the 10+10 Technology-to-Capability process.

	SCF #1:	SCF #2:	SCF #3:	SCF #4:	SCF #5:	SCF #6:	SCF #7:	SCF #8:	SCF #9:	SCF #10:	SCF #11:	SCF #12:
PCA #1:	✓	✓	✓	✓	✓	✓	✓	✓	✓	✓	✓	✓
PCA #2:	✓	✓	✓	✓	✓	✓	✓	✓	✓	✓	✓	✓
PCA #3:	✓	✓	✓	✓	✓	✓	✓	✓	✓	✓	✓	✓
PCA #4:	✓	✓	✓	✓	✓	✓	✓	✓	✓	✓		✓
PCA #5:	✓	✓	✓	✓	✓	✓	✓	✓	✓	✓	✓	✓
PCA #6:		✓	✓	✓	✓	✓	✓	✓	✓	✓		✓
PCA #7:	✓	✓	✓	✓	✓	✓	✓	✓	✓	✓	✓	
PCA #8:	✓	✓	✓	✓	✓		✓	✓	✓	✓		
PCA #9:	✓	✓		✓	✓	✓	✓	✓	✓	✓		
PCA #10:	✓	✓	✓	✓	✓	✓	✓	✓	✓	✓		
PCA #11:	✓	✓		✓	✓	✓	✓	✓	✓	✓	✓	
PCA #12:	✓	✓	✓	✓	✓		✓	✓		✓		
PCA #13:		✓		✓			✓	✓				
PCA #14:				✓			✓					
PCA #15:		✓			✓	✓	✓	✓				
PCA #16:						✓	✓			✓	✓	
PCA #17:						✓					✓	
PCA #18:		✓				✓		✓				
PCA #19:	✓	✓			✓	✓		✓		✓		
PCA #20:	✓	✓	✓			✓						
PCA #21:	✓	✓		✓	✓							
PCA #22:	✓	✓			✓							
PCA #23:		✓			✓		✓				✓	
PCA #24:		✓			✓		✓				✓	
PCA #25:					✓		✓					
PCA #26:			✓					✓				
PCA #27:			✓		✓		✓	✓				
PCA #28:			✓					✓				
PCA #29:			✓		✓			✓				
PCA #30:			✓								✓	

Figure 18. Mapping of PCAs across Air Force SCFs. (Courtesy of the Office of the Chief Scientist of the Air Force.)

Inherently Intrusion-Resilient Cyber Systems

This represents a fundamental shift in emphasis from "cyber protection" to "maintaining mission effectiveness" in the presence of cyber threats, using technologies such as Internet protocol (IP) hopping, network polymorphism, massive virtualization, and rapid network recomposition that can make cyber systems inherently resilient to intrusions entering through the network layer. These convert the currently static network layer into a highly dynamic one in which the hypervisor mapping between the hardware and functional layers changes constantly in a pseudorandom way, perhaps hundreds of times every second. A cyber adversary who finds vulnerabilities in the physical layer thus has virtually no time to use them for mapping the network before its topology has changed. Current manpower-intensive cyber protection efforts to block attacks at the network surface could be augmented or replaced by these new methods, which introduce "rapid maneuver" to cyber warfare to make a network inherently resilient to attack. Since adversaries entering or passing through the network layer may have only a few milliseconds to operate, not weeks or months, they leave greater forensic evidence for attribution. Intrusion-resilient methods would initially be applied to computer networks but are extendable to other cyber systems.

> *Commercial development of massive virtualization and hypervisor technologies to enable cloud computing can provide much of the technology needed for such inherently intrusion-resilient Air Force cyber systems.*

Automated Cyber Vulnerability Assessments and Reactions

Assuring continued mission effectiveness in the cyber environment begins with identifying mission essential functions across the Air Force enterprise and prioritizing these in terms of individual and interdependent mission impacts. Mapping the dependence of each function over the cyber domain gives insights into critical hidden interdependences that enable autonomic generation of appropriate reactions to attacks at the "speed of cyber" and thereby maximizes mission effectiveness under any threat. This is beginning to be done today via a slow manual process for limited and specific threats; entirely new automated approaches can allow scale-up across the entire

Air Force cyber enterprise and automatic development of optimal courses of action for any momentary threat set across the enterprise to maintain maximum mission effectiveness. Software tools to autonomously probe and map complex interdependences across the cyber enterprise are still in relative infancy, and tools needed to visualize and understand the resulting cyber vulnerabilities are primitive. Autonomous decision support tools will also be needed to allow reactions at speeds needed to limit the effects of cyber attacks. Such automated mapping of cyber vulnerabilities and corresponding attack responses represents one way of implementing the overarching theme of "resilience" in Air Force cyber systems.

> *Cyber defense must shift emphasis from network protection to maintaining enterprise-wide mission effectiveness; automated methods for continuous assessment of interdependences among cyber elements and generation of autonomic reactions to threats will be key.*

Decision-Quality Prediction of Behavior

Advances in human and cultural behavior modeling, social network modeling, cognitive modeling, and autonomous reasoning can enable decision support tools for anticipating and predicting adversary and own-force behaviors. Fusion and understanding of information from disparate sources, including classical intelligence data as well as large data sets of open source information routinely collected from global cyber networks, provide essential inputs. Massive analytics made possible by the availability of near-infinite processing and storage capacities during this time can provide software tools that aid in decision making, course-of-action development, and related tasks involving prediction of individual/collective behaviors over a range of conditions, along with statistical uncertainty bounds to quantify decision confidence levels and inform alternative courses of action.

> *Massive storage and processing power during this time will enable wider use of open source and other information and advanced statistical models of individual or collective behaviors.*

Augmentation of Human Performance

Human performance augmentation will be essential for effectively using the overwhelming amounts of data that will be routinely available during this time. As suggested in figure 19, this may include implants, drugs, or other augmentation approaches to improve memory, alertness, cognition, and visual/aural acuity. It may even extend to limited direct brainwave coupling between humans and machines and to screening of individual capacities for key specialty codes via brainwave patterns and genetic correlators. Adversaries may use genetic modification to enhance specific characteristics or abilities. ISR operators, commanders, and personnel in the cockpit and on the flight line will routinely use performance augmentation applications. Data may be fused and delivered to humans in ways that exploit synthetically augmented intuition to achieve needed decision speeds and enhance decision quality. Human senses, reasoning, and physical performance will be augmented using sensors, biotechnology, robotics, and computing power.

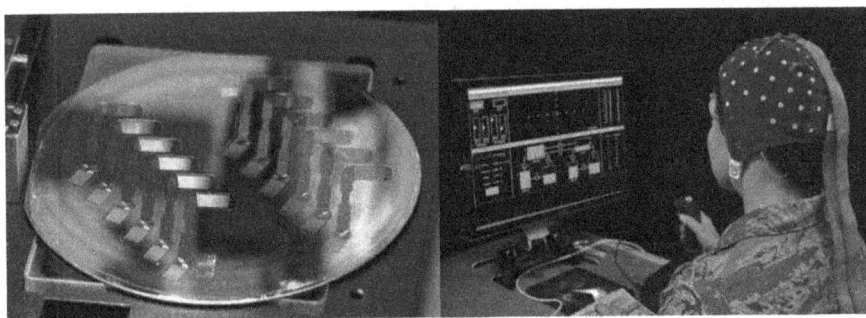

Figure 19. Technologies to augment human performance and cognition have significant potential for reducing Air Force manpower needs or increasing effectiveness in specialized roles. *Left,* electronics for a third-generation artificial retina that could restore sight to millions; *right,* augmentation of human performance through brainwave technology. (Courtesy LLNL, Public Affairs Release NR-10-02-03, "Lab Plays Key Role in Department of Energy's Artificial Retina Project," 4 February 2010 [*left*], and US Air Force [*right*].)

> *As natural human capacities become increasingly mismatched to data volumes, processing capabilities, and decision speeds, augmenting human performance will become essential for gaining the benefits that other technology advances can offer.*

Advanced Constructive Discovery and Training Environments

Constructive environments being used for training today are simple by the standards of commercial massively multiplayer online games (MMOG), many of which have hundreds of thousands of users around the world playing at any given moment in thousands of simultaneous multiplayer games; CounterStrike alone has over 4.5 billion man-minutes of game play worldwide every month. Advances in such highly constructive environments—compared with virtual systems that are in Air Force use today—enabled by continued processing and storage capacities can provide sufficient fidelity to allow useful mining of the enormous statistical information contained in such global game play. This can enable entirely new methods for discovering cultural insights as well as innovative blue- and red-force concepts of employment long before they are evident by ordinary experience or in the far smaller statistical samples available by formal methods. Highly constructive environments also enable further improved training at reduced cost and will be used for high-fidelity mission rehearsal, course-of-action development, and other uses (fig. 20).

Figure 20. Virtual and constructive technologies can be advanced to allow high-fidelity, constructive, massively multiplayer environments for discovery, including online environments where game play statistics are mined to discern cultural insights and identify potential concepts of employment. (Courtesy US Air Force.)

> The fidelity of constructive environments will benefit from commercial advances in processing, storage, and advanced authoring tools developed for online game markets; expanding the use of such environments offers substantial new capabilities as well as cost savings.

Trusted, Adaptive, Flexibly Autonomous Systems

Inexpensive processing power, data storage, and powerful autonomous reasoning algorithms will enable systems capable of far higher levels of autonomy and extension of autonomous systems to entirely new functions. Such advanced autonomous systems will be powerful force multipliers (as suggested in fig. 21) and will enable operations at timescales far faster than possible with human-in-the-loop control. Establishing trust in autonomy will thus become the central factor in gaining access to the potentially enormous capabilities that such systems can offer. Human control will be preserved with legally reviewed rules of engagement and appropriate human-in-the-loop decision points. However, adversaries are unlikely to be constrained by such limits and may be able to field systems that can take advantage of far greater levels of autonomy. Entirely new approaches will be needed to enable V&V for the near-infinite-state control systems created by high levels of adaptability and autonomy. Developing such methods will be essential for gaining certifiable trust and legal freedom to field autonomous systems that can match or exceed potential adversary capabilities.

> Methods to enable trust in highly adaptive autonomous systems are a game-changing technology needed to reap the enormous capability and cost benefits that such systems can offer and to avoid the asymmetric advantage that adversaries could otherwise have with such systems.

Frequency-Agile Spectrum Utilization

Future Air Force systems will need substantially frequency-agile capabilities to ensure access to available spectrum bands in worldwide operations. Spectral agility will also be needed for jam resistance, low probability of detection/intercept, and cyber resilience in the increasingly congested spectrum environment and increasingly contested EW environment. Yet even software-defined spectral agility has been diffi-

Figure 21. Manpower-intensive Air Force functions performed using trusted autonomous systems can enable substantial manpower efficiencies and provide time-domain operational advantages over adversaries limited to human planning and decision processes. (Courtesy US Air Force.)

cult to implement and certify. New methods such as dynamic spectrum access and associated policy engines will require compatibility with Air Force–specific aspects of spectrum use, including fast movers, large fields of regard, compatibility with below-noise signals such as radars, and resistance to malevolent interferers. Frequency agility can also be extended to spectral mutability, in which a wider range of spectral parameters may be adjustable.

> *Frequency agility will be essential for ensuring that Air Force systems can operate under rapidly changing spectrum regulations worldwide and can provide needed functionality in contested environments where electronic and cyber warfare could create cross-domain vulnerabilities.*

Dominant Spectrum Warfare Operations

Today the Air Force relies on US joint force partners for EW support, yet it will become an essential resident capability for many types of systems. Growing overlaps in electronic and cyber warfare extend this need to full-spectrum dominance, including electronic attack, defense and support, and corresponding functions for optical free-space and wave-guided EM systems. In modern IADS environments, jamming may have to overcome sophisticated electronic counter-countermeasures in radars and SAM seekers via methods such as cognitive EW. Increasing use of track fusion methods may give adversary systems improved jam resistance, and increased standoff ranges imposed by long-range SAM systems further complicate spectrum warfare. As potential adversaries increasingly move to AESA radars and make use of DRFM-based systems for RF spoofing, Air Force freedom of operation in the air-to-air EM domain is also being significantly challenged. New spectrum warfare methods should be developed to support Air Force dominance in attack, self-defense, and support across the EM spectrum.

> *Rapidly advancing EW capabilities of some potential adversary systems may create significant challenges to traditional Air Force spectrum dominance; a new generation of spectrum warfare methods is needed to ensure acceptable freedom of operations.*

Precision Navigation/Timing in GPS-Denied Environments

Widespread dependence of critical Air Force and other DOD systems on GPS for precision navigation and timing functions has caused potential adversaries to exploit GPS jamming as an asymmetric advantage. Key systems will thus need GPS independence or augmentation to allow their use in such environments. Chip-scale atomic clocks as shown in figure 22 and inertial measurement units (IMU) based on cold-atom principles or other technologies as shown in figure 23 can provide low-drift PNT in the event of GPS loss. These approaches maintain GPS-like position and timing uncertainties over relatively long periods after GPS signal loss. Netting to systems outside the jammed environment will allow relaying intermittent reference posi-

Figure 22. Chip-scale atomic clocks can provide low-drift timing independent of active GPS links, allowing a backup timing source to permit a wide range of critical Air Force systems to continue operating even under GPS-denied conditions. (Reprinted from National Institute of Standards and Technology [NIST], "NIST Unveils Chip-Scale Atomic Clock," 27 August 2004, accessed 20 June 2011, http://www.nist.gov/public_affairs/releases/miniclock.cfm.)

tion and timing information for limited updates to correct drift during long-duration jamming events. Technologies can enable miniaturization of such systems and supporting system-level network functions to negate the asymmetric adversary advantage that GPS jamming could otherwise provide.

> *Developing miniaturized cold-atom IMUs and clocks as well as other approaches for chip-scale PNT will be essential to ensure access to PNT in GPS-denied areas.*

Figure 23. Cold-atom devices enable low-drift IMUs and clocks for PNT in GPS-denied environments. Intersecting lasers trap/cool ytterbium, strontium, or other atoms to slow their motion, forming a Bose-Einstein condensate that can serve as a highly precise matter-wave interferometer. (Reprinted from NIST, "Experimental Atomic Clock Uses Ytterbium 'Pancakes,'" *NIST Tech Beat*, 6 Mar 2006, accessed 21 June 2011, http://www.nist.gov/public_affairs/techbeat/tb2006_0306.htm.)

Next-Generation High-Bandwidth Secure Communications

Laser communication technology allows extremely high-bandwidth communications over moderate ranges in clear weather conditions for atmospheric propagation, and over very long ranges when used for cross-links in space. The extremely low side lobes produced in transmissions at optical wavelengths also allow for potentially secure communications. When coupled with quantum key distribution (QKD), laser communication can provide verifiably secure encrypted transmissions. These characteristics make lasers attractive for linking critical

Air Force satellite, air, and ground network nodes and can help shift the rapidly growing demand for bandwidth to frequencies where spectrum management is far simpler. Developing precision pointing and other technologies to enable and integrate laser communications in key Air Force systems will be a high priority.

> Technologies to enable laser-based communication links with QKD methods can provide large increases in bandwidth and security needed in some types of Air Force uses.

Persistent Near-Space Communications Relays

Flight vehicles and lighter-than-air systems as in figure 24 can serve as pseudosatellites capable of operating at near-space altitudes as communication relays in the event of satellite loss. Netting several such relays can provide theater-level coverage and full connectivity to the continental United States, with system-level concept of operations providing needed platform protection. These systems hold significant platform challenges as well as electronics, power, and thermal management challenges but may be key to assured communications under the increasingly contested space environment. Systems designed to remain airborne for extremely long durations can allow launch and ascent to be timed when winds are sufficiently weak, enabling structural masses substantially below those of traditional airships designed for lower-altitude use. Significant technical challenges must be overcome in high-altitude materials, lightweight solar cells and energy storage, and thermal management to enable such systems.

> Near-space communications relays can provide essential backup for satellite communications in the increasingly contested space domain but represent substantial technical challenges.

Processing-Enabled Intelligent ISR Sensors

ISR sensors equipped with backplane processing for data synthesis and fusion can permit certain PED functions to be performed on the sensor itself, reducing bandwidth otherwise consumed in transferring large amounts of raw data to the ground. Functions such as coherent

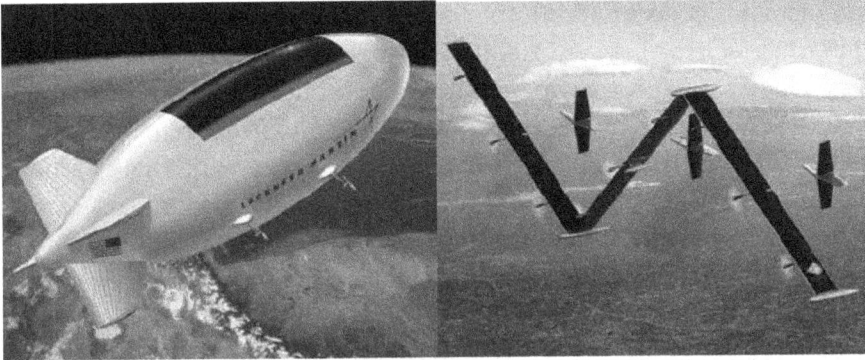

Figure 24. Persistent near-space systems, potentially in the form of ultra-long-endurance airships or autonomous flight vehicles, can ensure theater-level communications relay functions in the event of loss or degradation of corresponding space systems. (Courtesy Lockheed Martin [*left*] and "DARPA's Vulture: What Goes Up, Needn't Come Down," *Defense Industry Daily*, 16 September 2010, http://www.defenseindustry daily.com/DARPAs-Vulture-What-Goes-Up-Neednt-Come-Down -04852 [*right*].)

change detection may be suited to on-sensor processing, with raw data being transmitted only when cued by processed results from the sensor. On-sensor processing is enabled by the massive processing and storage capacity that can be routinely integrated into modern electronic systems. Processing done on the sensor does not need to be as detailed as can be done on the ground, since its role is only to provide initial cueing that triggers raw data transmission for the ground processing. Beyond substantial reductions in bandwidth consumed by transmission of raw sensor data, intelligent sensors can greatly reduce the manpower needs of current ground-based PED of sensor data.

> *On-sensor processing of raw data to cue transmission and detailed off-board processing can greatly reduce bandwidth consumed by ISR platforms and reduce PED manpower needs.*

High-Altitude, Long-Endurance ISR Airships

Theater-level ISR aircraft operating for long durations in fixed orbits require forward flight to generate aerodynamic lift, imposing fuel and maintenance costs in addition to engine and aircraft structural masses that cause sensor payload fractions to below those of airships. HALE airships designed for extremely long endurance can achieve payload mass fractions even higher than those of traditional airships, since their ascent through the atmosphere can be timed when winds aloft are low. The size of HALE airships, as shown in figure 25, accommodates massive sensor apertures far larger than any other aircraft could carry, enabling extreme RF sensitivities that may be essential in the emerging strategic environment. Their high operating altitude provides a large field of regard, allowing such systems to also serve as theater-level near-space communication relays. Technical challenges that must be overcome include advancement of high-altitude, long-life materials; lightweight solar cells and energy storage; and lightweight multifunctional sensor structures.

> *HALE airships can provide theater-level ISR capabilities with massive sensors and offer inherently low operating costs compared with traditional aircraft-based systems.*

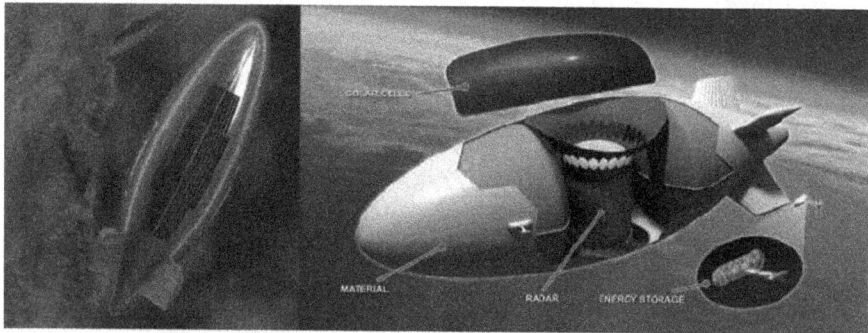

Figure 25. HALE ISR airships can remain aloft for years so that launch/ascent is decoupled from immediate demand and thus can be timed when winds are low. This allows exceptionally lightweight structures that can enable extreme functionality to be used. (Courtesy Lockheed Martin, reprinted from *Wikipedia*, s.v. "High-Altitude Airship," accessed 20 June 2011, http://en.wikipedia.org/wiki/High_altitude_airship [*left*], and DARPA [*right*].)

Prompt Theater-Range ISR/Strike Systems

Air-breathing hypersonic systems designed for Mach 6 or below avoid most of the extreme temperatures traditionally associated with hypersonic flight, yet can provide rapid ISR or strike capability to engage high-value, time-critical targets from standoff distances. Such systems can hold targets at risk that might otherwise be immune, potentially altering the calculus of strategic deterrence. The system in figure 26, for example, uses combined-cycle propulsion with internal rocket boost, ram/scramjet acceleration, and scramjet cruise. Inward turning inlets and a dual-flow path design allow high volumetric efficiency, and high cruise speed provides significantly increased survivability. Vertical takeoff using a rocket-based combined cycle gives substantially greater range and payload while imposing few real limitations on launch sites relative to turbine-based horizontal takeoff systems.

> *High-speed theater-range ISR/strike systems can provide time-critical mission capabilities in a reusable system suited to ranges and operating conditions of key strategic missions.*

Fractionated, Survivable, Remotely Piloted Systems

A standardized, low-cost, small- to medium-sized, remotely piloted airframe able to carry mission-specific payloads can enable a fractionated and survivable approach for meeting key needs of missions spanning the spectrum from low-end to high-end operations. An as-needed mission package can be composed from such standard elements carrying different payloads, ranging from single or paired elements for uncontested environments to a dozen or more elements in A2/AD environments. The underlying airframe that enables such an architecture would use small, efficient turbojet propulsion; be capable of fully autonomous takeoff, flight, landing, and collaborative control among elements of the mission package; and would carry standard-sized modular payloads such as ISR, EW, strike, communications, and other functions among the individual elements to compose the mission capability. Elements cooperate autonomously to form a coherent system using secure, burst-mode, frequency-agile RF or laser communication. Beyond making use of affordable LO technologies, mission survivability

Wall Temperature, TWALL (K)

589.302856
659.568054
729.833252
800.098450
870.363647
940.628906
1010.8941
1081.1593
1151.4246
1221.6897
1291.9550
1362.2201
1432.4854
1502.7506
1573.0157
1643.2809

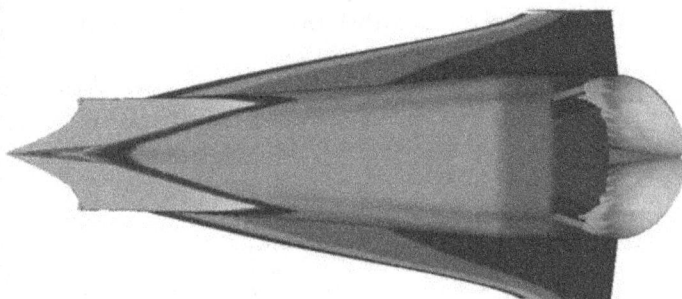

Mach Number

1.437533
2.135761
2.833989
3.532218
4.230446
4.928674
5.626902
6.325130
7.023358
7.721587
8.419815
9.118043
9.816272
10.514500
11.212729
11.910956

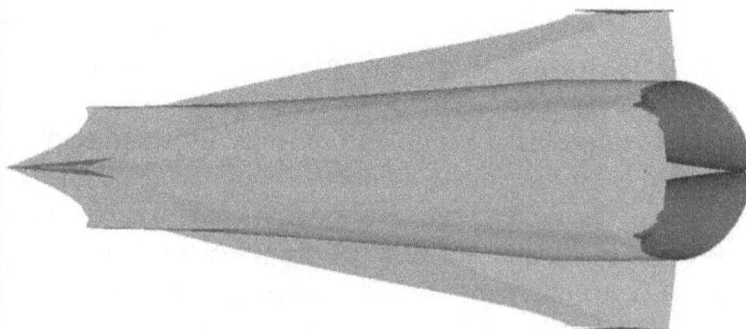

Figure 26. Single-stage ISR/strike vehicles such as this early-stage concept design, which uses internal rocket boost to Mach 3.5 and rocket-based combined-cycle acceleration to scramjet cruise at Mach 6, can enable time-responsive missions at long ranges with substantial payloads and runway landings. (Courtesy Astrox Corporation.)

in A2/AD environments would come from the multiplicative benefits of redundancy among the low-cost expendable elements in the fractionated system architecture, which can overwhelm adversary IADSs via fundamentally asymmetric cost-imposing strategy.

> *Fractionated autonomous systems provide a low-cost means for as-needed composable mission capabilities, from individual or paired systems for simple EO-IR imaging or hunter-killer missions to large mission packages with redundancy for A2/AD environments.*

Direct Forward Air Delivery and Resupply

Small-unit operations in difficult environments are likely to require improved means for resupply beyond current joint precision airdrop system (JPADS) capabilities and related near-term systems. Such a capability will typically need substantial autonomy and aerodynamic control authority to overcome high winds in mountainous terrain while maintaining precise drop-point accuracy, requiring substantially different aerodynamic and control concepts than traditional autonomous systems. Demand for a wide range of weight classes and the ability to deliver from existing mobility platforms make this a substantially challenging system.

> *The likelihood of a continuing mid- to long-term need to provide precision forward air delivery in difficult terrain makes this an important challenge for advancing autonomous system capabilities within significant limits on cost and other factors.*

Energy-Efficient, Partially Buoyant Cargo Airlifters

Partially buoyant cargo airlifters as shown in figure 27 derive most of their lift from buoyancy and the remainder aerodynamically from forward flight. They typically differ from dirigibles in that they have a rigid semimonocoque hull structure fabricated from a lightweight but strong composite that provides significant structural mass efficiencies. Moderate-size prototypes of such craft have been built and operated, and fully autonomous systems are possible. Large systems of this type can transport enormous amounts of cargo far more energy efficiently than winged aircraft and more quickly than traditional

Figure 27. Hybrid airships that achieve part of their lift from buoyancy and part aerodynamically from forward flight can serve as highly efficient large-scale cargo airlifters for long-haul routes and in uncontested areas with unprepared landing sites. (Courtesy Lockheed Martin.)

seagoing cargo ships, and can land in relatively unprepared sites to deliver cargo closer to the point of use. Enabling such capabilities requires advancing several of the underlying technologies in system design and integration for differing weight classes and operating capabilities, as well as in technologies to address weather effects, ballast control, and cargo handling.

> *Partially buoyant cargo airlifters could provide substantially reduced fuel costs for certain types of air mobility missions and increased ability to deliver large-scale cargo into relatively unprepared sites in a potentially autonomous system.*

Fuel-Efficient Hybrid Wing-Body Aircraft

Subsonic hybrid wing-body aircraft provide significantly increased fuel efficiency and are adaptable for multiple mission roles. Configurations such as in figure 28 are enabled in part by advanced lightweight

Figure 28. Hybrid wing-body aircraft have higher fuel efficiency than traditional tube-and-wing designs for tankers and transports. Advances in integrated stitched-composite manufacturing technologies such as PRSEUS (pultruded, rod-stitched efficient unitized structure) have enabled large, lightweight, high-stiffness panels at low cost. (Courtesy NASA/Dryden.)

stitched-composite fabrication methods that allow out-of-autoclave forming of large, complex-shaped panels. Hybrid wing-body designs have a wide internal bay suitable for cargo hauling and personnel transport or for use as aerial-refueling tankers. A single common airframe of this type could be reconfigured during its service life to any of these mission sets to meet changing Air Force mission needs. These designs are suitable to be developed into a family of related aircraft with common wing and other components that can potentially enable reduced production and maintenance costs.

> *Hybrid wing-body aircraft fabrication is enabled by PRSEUS technology that can substantially reduce airframe weight and cost.*

Next-Generation High-Efficiency Turbine Engines

Demands for reducing fuel costs and increasing range and endurance call for substantially increased efficiency of embedded turbine engines. Continuing advances in component-level and system-level methods to achieve lower fuel consumption will be needed beyond efforts currently under way, as in figure 29. These are driving overall pressure ratios (OPR) well above today's state of the art for increased Brayton cycle efficiencies, in turn requiring new advances in materials and thermal management technologies. At the same time, emerging "third-stream engine architectures" can enable constant-mass-flow engines that can provide further reductions in fuel consumption. A wide range of technology advances can enable substantial collective improvements in turbine engine fuel efficiency.

> The Air Force's need for high-efficiency embedded turbine engines in many of its platforms calls for substantially different approaches from those accessible to commercial high-bypass engines.

Embedded Diagnostic/Prognostic Subsystems

An embedded system of diagnostic and prognostic sensors emplaced within high-value Air Force assets such as air vehicles and DE systems will provide warning and prediction of structural or subsystem failure.

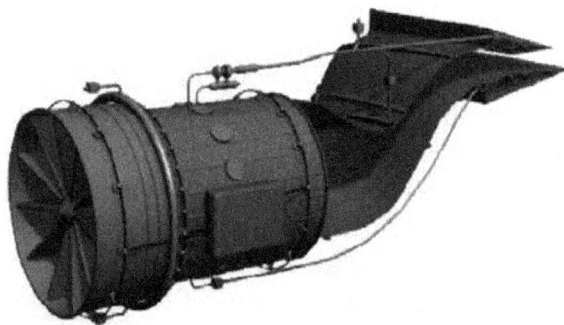

Figure 29. Advancing and integrating technologies that enable a generation of highly efficient embedded turbine engines at a range of scales will be key for meeting numerous Air Force needs, from fuel cost savings to long-endurance ISR and long-range strike. (Courtesy US Air Force.)

These systems will ensure that preventive maintenance is performed when actually necessary rather than by antiquated time-based maintenance schedules. Data will be autonomously tracked by individual tail/system number and part by part. The system can alert maintenance personnel of suspect parts and queue the supply system to ensure replacement prior to failure. These subsystems can also allow maintenance personnel to impose operational flight-envelope restrictions on individual tail numbers as needed, rather than imposing these across the entire fleet. Such embedded subsystems can increase overall operational readiness while decreasing maintenance costs.

> *Embedded sensors and processors to determine system health and maintenance requirements can extend service life of many systems and reduce maintenance and sustainability costs.*

Penetrating, Persistent Long-Range Strike

The need to penetrate substantially defended adversary airspace and persist to achieve desired effects is likely to remain an enduring requirement. This will demand a range of technologies that can achieve deep penetration capability, long loiter, and secure egress. Such systems would likely need LO across multiple bands, secure communications, and advanced active and passive spectrum warfare capabilities to allow deep penetration into adversary territory. Technologies to support efficient engines, bleedless inlets, and serpentine nozzles will be needed for improved propulsion integration to enable long range and loiter.

> *Maintaining LO capabilities is likely to remain essential for key Air Force missions, as will a wide range of supporting technologies needed to enable long range and persistence.*

High-Speed Penetrating Cruise Missile

A high-speed cruise missile as suggested in figures 30 and 31 can provide an effective standoff component in a long-range strike system. It achieves penetration into A2/AD environments by using speed to obtain substantially increased survivability against modern air defenses. Such an air-breathing penetrating cruise missile might operate in the

Figure 30. Advances made over the past decade in scramjet propulsion technologies embodied in the X-51 demonstrator can allow development of high-speed cruise missiles and other capabilities to support long-range strike on critical targets in A2/AD environments. (Courtesy US Air Force.)

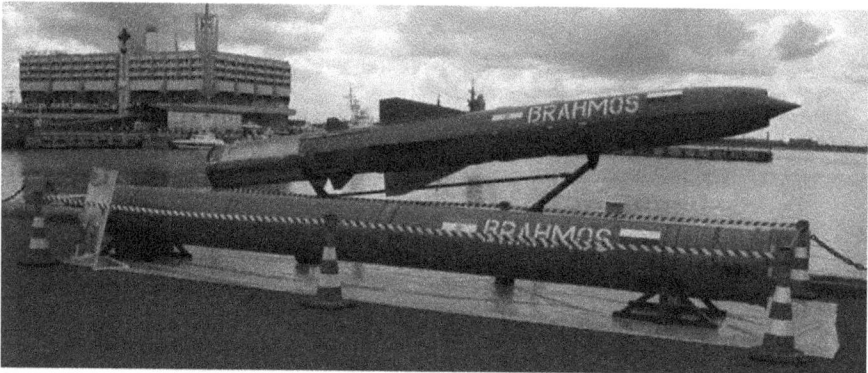

Figure 31. India and Russia have developed and fielded the BrahMos supersonic cruise missile, reportedly capable of Mach 2.5 flight over ranges in excess of 250 kilometers. Su-30 fighters are reportedly being modified to provide an air-launch capability, primarily for antiship and land attack roles. (Reprinted from *Wikipedia*, s.v. "BrahMos," accessed 20 July 2011, http://en.wikipedia.org/wiki/ Brahmos.)

high supersonic or low hypersonic regime to provide a balance between flight time, range, and survivability. Advances in dual-mode ram/scramjet propulsion can enable systems up to around Mach 6, while high Mach turbine engine advances would be used for systems operating closer to Mach 3+. Such a system might be air-launched and employ multiple redundant systems for navigation and target acquisition including GPS, chip-scale cold-atom PNT systems, optical terrain recognition, EO and IR sensors, passive RF sensing, and active EW. Variants could include a hard target penetrator or dispenser for submunitions.

> Speed provides an alternative to low observability that can permit time-critical strikes in A2/AD environments.

Hyperprecision Low-Collateral-Damage Munitions

The development of focused lethality munitions, precision guidance via cold-atom PNT systems, and increased ISR information allows the manufacture of intelligent, cooperative, small munitions with near-zero circular error probable (CEP) capable of identifying and coordinating an attack through simultaneous engagements of a target to achieve a mission-defined level of effect. Multiple munitions could be launched or dropped from one or several air vehicles to cooperatively engage targets. Operating autonomously or with a continuous stream of ISR data, the munitions would do final target acquisition and cooperatively plan a simultaneous attack to achieve the desired effect. Munitions not used would engage other approved targets or self-destruct with minimal collateral damage.

> Advances in precision guidance for GPS-denied areas and focused-lethality munitions can enable a generation of hyperprecise, low-collateral-damage munitions.

Directed Energy for Tactical Strike/Defense

Laser and high-power microwave systems can be developed with sufficient power and beam control for active air base defense, air vehicle defense, and tactical strike applications, as suggested in figure 32. Prior chemical lasers as used in the advanced tactical laser will be replaced by solid-state laser systems, most likely derived from the High-Energy Liquid Laser Area Defense System (HELLADS) technology and subsequently will be replaced by even more efficient fiber laser systems. The reduced size, weight, power, and thermal management requirements will allow closer integration of such systems in advanced fighters and

Figure 32. Laser DE systems will use advanced solid state and fiber laser technologies that enable lightweight, efficient airborne self-defense and low-collateral-damage tactical strike. (Courtesy US Air Force.)

other tactical platforms. Highly autonomous, self-contained systems can be utilized for self-defense to neutralize credible GPS receiver application module (GRAM) and AAM missile threats. Airborne tactical strike systems can be used against targets in complex environments such as urban areas with near-zero collateral damage. Such systems can incorporate intelligent beam control to avoid collateral damage to blue forces and noncombatants. Advances in power and thermal management can permit ground-based air base defensive systems to have near-infinite magazines, while airborne systems will have magazines limited by onboard power generation and storage and by thermal management capabilities.

> *Technologies for laser-based DE systems enabling near-zero-collateral-damage tactical strikes and airborne self-defense are advancing rapidly and in the midterm to long term may allow such systems to expand into new airborne and ground-based roles.*

Enhanced Underground Strike with Conventional Munitions

Hyperprecision navigation, timing, and autonomous/cooperative guidance will enable two or more conventional munitions to strike high-value underground targets using extremely short and precisely spaced time separations to provide maximal penetration. Using the unsteady dynamics of the compression and expansion wave system produced in a granular earth medium by the downward impulse from a preceding munition allows a following time-phased munition to penetrate through the locally weakened depth-time region produced by the overexpansion wave in the resulting wave system. In effect, the second munition "wave rides" through the subsurface wave system created by the first munition, taking advantage of the degraded material properties of the granular medium that result from partial dislocation of the otherwise interlocking granular constituents. This partial fluidization of the granular medium allows for dramatically reduced drag on the second munition as well as increased penetration depth and stability.

Relatively soft underground targets can be struck without ground-penetrating munitions by using precisely time-phased impacts that allow subsurface wave dynamics to be exploited for achieving substantially deeper impact than would otherwise be possible.

Reusable Air-Breathing Access-to-Space Launch

The cost of placing satellites in orbit with current EELV launchers is high. Next-generation two-stage-to-orbit (TSTO) launch systems currently under consideration are focused on reusable rocket approaches that can provide modest reductions in launch costs, depending on launch rates. However, reusable air-breathing TSTO systems based on combined-cycle propulsion have potential to create large reductions in launch costs since such systems utilize oxygen in the atmosphere and thus do not need to loft the enormous oxidizer weight needed by rocket systems. Vertical takeoff space launch vehicles such as this will use a reusable rocket in the first stage with a reusable rocket-based, combined-cycle air-breathing scramjet in the second stage. Advanced materials, automation, and onboard diagnostic and prognostic systems will allow rapid turnaround for reuse as well as life cycle cost savings as compared to conventional rockets.

Air-breathing, combined-cycle TSTO launch systems have the ability to provide substantial reductions in space launch costs.

Rapidly Composable Small Satellites

Small satellites, as in figure 33, can be assembled, tested, launched, and functioning on orbit within days of operational requirement generation. These satellites are based on a plug-and-play/open-architecture approach using standards for self-describing components within a discoverable and autoconfiguring system. Satellite control can be accomplished with standardized ground stations. These architectures can be sufficiently flexible to allow spiral upgrade of individual components,

Figure 33. Small satellites composed from standardized modular elements can provide rapid on-orbit capabilities with "good enough" functionality. Fractionated or distributed space systems composed from such small satellites can significantly increase their functionality and survivability. (Courtesy NASA.)

cooperative control with other satellites, and the potential for on-orbit self-assembly. Subassemblies can include propulsion for attitude control, orbital maintenance and maneuver, laser and/or traditional communications, ISR, SSA, self-defense, and other missions. Such plug-and-play architectures are based on a cross-domain-enabling concept since air and cyber systems can also be built from preexisting modular components using similar "open architecture" approaches.

> *Plug-and-play modular components allow lower-cost small satellites with "good enough" capabilities to be rapidly designed and assembled for operationally responsive space missions.*

Fractionated/Distributed Space Systems

Constellations of smaller satellites operating cooperatively to perform a given mission set can provide greater survivability and greater ease of systematic upgrade. Architectures based on fractionation involve several functionally different satellite elements that cooperate to act as a single system, while distributed architectures involve many copies of identical elements that operate coherently to produce greater capability than each element could individually. The individual satellite elements make use of onboard data processing to minimize interconstellation communications to achieve a highly secure and jam-resistant architecture. Such systems may utilize plug-and-play/open architectures and allow seamless upgrade of the constellation's overall capability by adding or substituting new individual, small satellites.

> *Fractionated or distributed space systems can potentially provide greater on-orbit capability at lower cost than is achievable with traditional approaches, while allowing rapid reconstitution in the event that elements In the constellation are lost.*

Persistent Space Situational Awareness

Current ground-based radars and telescopes and space-based space surveillance (SBSS) assets that comprise the space surveillance network can be augmented to provide birth-to-death detection, tracking, and characterization of every object in orbit, from traditional large satellites to picosatellites and orbital debris at LEO, medium Earth orbit (MEO), and geosynchronous Earth orbit (GEO) altitudes. As suggested in figure 34, this can be done through a combination of new ground-based and space-based assets with appropriate fusion of data from other satellites, Aegis ships, and other sources into an integrated database. Current plans are to augment the radars and optical telescopes in the current space surveillance network with a 3.5-millimeter space surveillance telescope (SST), the SBSS system, and the S-band Space Fence system. These will address detection and tracking, but determining the contents of a satellite or its potential capabilities and intent will be extremely difficult. Bringing together data from active and passive RF and EO-IR sources can provide a true "SSA network" with capabilities far beyond those of its individual elements. In principle, all satellites in orbit can provide information that, when fused and analyzed, provides a far more complete picture, including space weather effects to allow discrimination of hostile actions from natural causes.

A substantially greater SSA capability will be essential to protecting space assets and allowing continued effective use of space.

Figure 34. Improved space surveillance capability will consist of ground-based and space-based assets that can provide birth-to-death tracking and characterization of all spaceborne systems. Effective data fusion from multiple sources provides significant leverage over single-source observations. (Courtesy US Air Force.)

Improved Orbital Conjunction Prediction

Development of a physics-based, near-real-time predictive model of near-Earth space environment interactions with spacecraft will allow highly accurate, actionable predictions of space object conjunctions. This model would utilize space weather data from ground- and space-based space environment sensors, SSA sensor data of object locations/trajectories, and substantially improved astrodynamic calculations using improved satellite drag models. The model will provide confidence levels in conjunction predictions sufficient to justify orbital maneuvers to reduce risks of collisions.

> *Development of such an improved conjunction prediction capability will be essential for protecting space assets from collisions and from resulting increases in orbital debris that would endanger critical Air Force space assets.*

Key Potential Capability Areas

The set of 30 PCAs is within reach as new Air Force capabilities by the 2030 target date of *Technology Horizons* and is matched to key needs of the strategic environment of this period. Among these PCAs, the following 12 were determined to be of highest priority:

- PCA1: Inherently Intrusion-Resilient Cyber Systems
- PCA2: Automated Cyber Vulnerability Assessments and Reactions
- PCA4: Augmentation of Human Performance
- PCA6: Trusted, Adaptive, Flexibly Autonomous Systems
- PCA8: Dominant Spectrum Warfare Operations
- PCA9: Precision Navigation/Timing in GPS-Denied Environments
- PCA12: Processing-Enabled Intelligent ISR Sensors
- PCA15: Fractionated, Survivable, Remotely Piloted Systems
- PCA19: Next-Generation High-Efficiency Turbine Engines
- PCA24: DE for Tactical Strike/Defense
- PCA27: Rapidly Composable Small Satellites
- PCA29: Persistent SSA

From Potential Capability Areas to
Key Technology Areas

The next section lists the KTAs that can enable each of the 30 PCAs that have been identified. In so doing, it identifies the S&T that must be pursued over the next decade to allow these new Air Force capabilities to be developed. Correlating the resulting lists of KTAs across all 30 PCAs, and particularly across the top 12 PCAs, makes it possible to determine the S&T investments that can make the greatest contributions to Air Force capabilities for the 2010–30 strategic environment.

Chapter 6

Key Technology Areas 2010–30

This chapter identifies a set of KTAs—among the most essential Air Force S&T to be pursued over the next decade—that will enable technology-derived capabilities that can respond to the strategic, budget, and technology environments.

Each of the PCAs noted in the previous chapter is predicated on a set of underlying technologies that enable materiel or nonmateriel means for achieving capabilities within that area. As indicated in the previous chapter (see fig. 17), the set of individual technology research efforts associated with these capabilities may be quite large; however, these individual technologies can be sensibly aggregated into broader technology areas.

Individual research efforts within any given technology area may be exploring different ways of achieving the same technology objective or may be exploring different technology objectives that are needed to enable complementary aspects of a given technology. It is not the objective of *Technology Horizons* to dictate which of these individual *lines* of research within a given technology area will likely prove to be the most productive. Rather, this effort is appropriately focused on identifying which technology *areas* are most essential for Air Force S&T to pursue to enable PCAs, such as those in chapter 5, that are aligned with the strategic environment, enduring realities, and overarching themes identified in chapters 2–4.

Focusing on such KTAs and not on individual lines of research is the appropriate level of specificity for an S&T vision. Identifying KTAs provides useful guidance in making difficult choices among the limited set of S&T investments that can be pursued in the likely budget environment over the next two decades, yet it gives AFRL leadership the freedom to evaluate and select the specific individual lines of research that should be pursued.

Key Technology Areas Supporting
Potential Capability Areas

For each of the PCAs identified in chapter 5, a large number of supporting technologies are needed to allow systems of that type to be developed or to contribute to the increased effectiveness of such systems. It has not been an objective of *Technology Horizons* to produce an exhaustive list of all such supporting technologies. Indeed, by being comprehensive, such a list would be of little or no use in providing guidance for Air Force S&T decisions.

Instead, this effort has sought to identify the more limited set of KTAs that not only support these PCAs but also are broadly essential for enabling the overarching themes previously noted (chap. 4, fig. 16). The resulting KTAs identified below for each of the PCAs should be understood in that way. They do not define the entirety of S&T that the Air Force needs, but they are among the most essential components of the S&T that the Air Force will need for the strategic environment, enduring realities, and overarching themes that define the 2010–30 time frame and beyond.

> A set of 110 individual technology areas is identified here as being among the most essential S&T for the Air Force to pursue over the next decade. Each of these KTAs is mapped into the PCAs from chapter 5 and the overarching themes from chapter 4.

Figure 35. Basic research is providing the foundation on which a wide range of future capabilities can be based, such as (*left*) superconducting artificial atom structures visualized via (*right*) amplitude spectroscopy and metamaterials with negative refractive index that bend EM waves backwards. (Courtesy US Air Force [*left*] and NASA [*right*].)

PCA1: Inherently Intrusion-Resilient Cyber Systems

- Ad hoc networks
- Virtual machine architectures
- Agile hypervisors
- Polymorphic networks
- Agile networks
- Pseudorandom network recomposition
- Laser communications
- Secure RF links
- Frequency-agile RF systems
- Spectral mutability
- Dynamic spectrum access
- QKD
- Complex adaptive distributed networks
- Complex adaptive systems
- Complex system dynamics
- V&V for complex adaptive systems

- Autonomous systems
- Autonomous reasoning
- Resilient autonomy
- Collaborative/cooperative control
- Decision support tools
- Automated software generation
- Distributed sensing networks
- Sensor data fusion
- Signal identification and recognition
- Cyber offense
- Cyber defense
- Cyber resilience
- Advanced computing architectures
- Complex environment visualization
- Massive analytics
- Automated reasoning and learning

PCA2: Automated Cyber Vulnerability Assessments and Reactions

- Ad hoc networks
- Polymorphic networks
- Agile networks
- Complex environment visualization
- Complex adaptive distributed networks
- Frequency-agile RF systems
- Spectral mutability
- Dynamic spectrum access
- QKD
- Complex adaptive systems
- Complex system dynamics
- V&V for complex adaptive systems
- Autonomous systems
- Autonomous reasoning and learning

- Resilient autonomy
- Collaborative/cooperative control
- Distributed sensing networks
- Decision support tools
- Information fusion and understanding
- Cyber defense
- Cyber resilience
- Human behavior modeling
- Cultural behavior modeling
- Social network modeling
- Complex environment visualization
- Massive analytics
- Human-machine interfaces

PCA3: Decision-Quality Prediction of Behavior

- Complex adaptive systems
- Complex system dynamics
- V&V for complex adaptive systems
- Autonomous systems
- Autonomous reasoning and learning
- Resilient autonomy
- Collaborative/cooperative control
- Decision support tools
- Distributed sensing networks
- Integrated sensing and processing

- Signal identification and recognition
- Information fusion and understanding
- Biological signatures
- Human behavior modeling
- Cultural behavior modeling
- Social network modeling
- Behavior prediction and anticipation
- Influence measures
- Cognitive modeling
- Human-machine interfaces

PCA4: Augmentation of Human Performance

- Chip-scale atomic clocks
- Ad hoc networks
- Virtual machine architectures
- Agile networks
- Complex adaptive distributed networks
- Complex adaptive systems
- Complex system dynamics
- V&V for complex adaptive systems
- Validation support to simulations
- Autonomous systems
- Autonomous reasoning and learning
- Resilient autonomy
- Collaborative/cooperative control
- Autonomous mission planning
- Health monitoring and prognosis
- Decision support tools
- Automated software generation
- Distributed sensing networks
- Integrated sensing and processing
- Sensor-based processing
- Signal identification and recognition
- Information fusion and understanding
- Advanced computing architectures
- Biological signatures
- Human behavior modeling
- Cultural behavior modeling
- Social network modeling
- Behavior prediction and anticipation
- Influence measures
- Cognitive modeling
- Complex environment visualization
- Massive analytics
- Cognitive performance augmentation
- Physical performance augmentation
- Human-machine interfaces

PCA5: Advanced Constructive Discovery and Training Environments

- Ad hoc networks
- Virtual machine architectures
- Agile networks
- Complex adaptive distributed networks
- Complex adaptive systems
- Complex system dynamics
- V&V for complex adaptive systems
- Multiscale simulation technologies
- Coupled multiphysics simulations
- Validation support to simulations
- Autonomous systems
- Autonomous reasoning and learning
- Resilient autonomy
- Collaborative/cooperative control
- Autonomous mission planning
- Decision support tools
- Automated software generation
- Signal identification and recognition
- Information fusion and understanding
- Cyber resilience
- Advanced computing architectures
- Human behavior modeling
- Cultural behavior modeling
- Social network modeling
- Behavior prediction and anticipation
- Influence measures
- Cognitive modeling
- Complex environment visualization
- Massive analytics
- Human-machine interfaces

PCA6: Trusted, Adaptive, Flexibly Autonomous Systems

- Complex adaptive systems
- Complex system dynamics
- V&V for complex adaptive systems
- Cold-atom inertial navigation systems (INS)
- Chip-scale atomic clocks
- Ad hoc networks
- Polymorphic networks
- Virtual machine architectures
- Agile networks
- Complex adaptive distributed networks
- Multiscale simulation technologies

- Validation support to simulations
- Autonomous systems
- Autonomous reasoning and learning
- Resilient autonomy
- Collaborative/cooperative control
- Autonomous mission planning
- Embedded diagnostics
- Health monitoring and prognosis
- Decision support tools
- Automated software generation
- Distributed sensing networks
- Integrated sensing and processing
- Sensor-based processing
- Signal identification and recognition
- Information fusion and understanding
- Cyber defense
- Cyber resilience
- Advanced computing architectures
- Human behavior modeling
- Cultural behavior modeling
- Social network modeling
- Behavior prediction and anticipation
- Massive analytics
- Human-machine interfaces

PCA7: Frequency-Agile Spectrum Utilization

- Chip-scale atomic clocks
- Ad hoc networks
- Polymorphic networks
- Agile networks
- Pseudorandom network recomposition
- Complex adaptive distributed networks
- Laser communications
- Short-range secure RF communications
- Frequency-agile RF systems
- Spectral mutability
- Dynamic spectrum access
- QKD
- Complex adaptive systems
- Complex system dynamics
- V&V for complex adaptive systems
- Autonomous systems
- Autonomous reasoning and learning
- Resilient autonomy
- Collaborative/cooperative control
- Autonomous mission planning
- Decision support tools
- Advanced RF apertures
- Secure RF links
- RF EW
- Distributed sensing networks
- Integrated sensing and processing
- Signal identification and recognition
- Information fusion and understanding
- Cyber resilience
- Advanced computing architectures
- Complex environment visualization
- Massive analytics
- Human-machine interfaces
- Optical and infrared materials
- RF and electronic materials
- Metamaterials

PCA8: Dominant Spectrum Warfare Operations

- Cold-atom INSs
- Chip-scale atomic clocks
- Ad hoc networks
- Polymorphic networks
- Virtual machine architectures
- Agile networks
- Pseudorandom network recomposition
- Complex adaptive distributed networks
- Modular small-sat components
- Distributed small-sat architectures
- Fractionated small-sat architectures
- Laser communications
- Short-range secure RF communications
- Frequency-agile RF systems
- Spectral mutability
- Dynamic spectrum access
- QKD
- Complex adaptive systems
- Complex system dynamics
- V&V for complex adaptive systems
- DE effects
- DE protection

- High-power microwaves
- Lightweight multifunctional structures
- Autonomous systems
- Autonomous reasoning and learning
- Resilient autonomy
- Collaborative/cooperative control
- Autonomous mission planning
- Decision support tools
- Passive radar
- Advanced RF apertures
- Secure RF links
- RF EW
- EO-IR sensing
- IR signature suppression
- Distributed sensing networks
- Integrated sensing and processing
- Sensor-based processing
- Signal identification and recognition
- Information fusion and understanding
- Cyber offense
- Cyber defense
- Cyber resilience
- Optical and infrared materials
- RF and electronic materials
- Metamaterials
- Nanomaterials

PCA9: Precision Navigation/Timing in GPS-Denied Environments

- Cold-atom INSs chip-scale atomic clocks
- Ad hoc networks
- Polymorphic networks
- Agile networks
- Laser communications
- Short-range secure RF communications
- Frequency-agile RF systems
- Spectral mutability
- QKD
- Complex adaptive systems
- Complex system dynamics
- V&V for complex adaptive systems
- Autonomous systems
- Autonomous reasoning and learning
- Resilient autonomy
- Collaborative/cooperative control
- Advanced RF apertures
- Secure RF links

PCA10: Next-Generation High-Bandwidth Secure Communications

- Cold-atom INSs
- Chip-scale atomic clocks
- Ad hoc networks
- Laser communications
- Short-range secure RF communications
- Frequency-agile RF systems
- Spectral mutability
- Dynamic spectrum access
- QKD
- Solid-state lasers
- Fiber lasers
- Semiconductor lasers
- EO-IR sensing
- Optical and infrared materials
- RF and electronic materials
- Metamaterials

PCA11: Persistent Near-Space Communications Relays

- Advanced aerodynamic configurations
- Aerodynamic test and evaluation (T&E)
- Cold-atom INSs
- Chip-scale atomic clocks
- Ad hoc networks
- Polymorphic networks
- Agile networks
- Laser communications
- Short-range secure RF communications
- Frequency-agile RF systems
- Spectral mutability
- Dynamic spectrum access
- DE protection
- Lightweight multifunctional structures
- Advanced composite fabrication
- Structural modeling and simulation (M&S)
- Multiscale simulation technologies
- Coupled multiphysics simulations

- Validation support to simulations
- Embedded diagnostics
- Health monitoring and prognosis
- High-altitude airship
- Advanced RF apertures
- Secure RF links
- RF EW
- Cyber defense
- Cyber resilience
- High-altitude materials
- Lightweight materials
- Advanced composites

- Composites sustainment
- RF and electronic materials
- Metamaterials
- Self-healing materials
- Nanomaterials
- Material-specific manufacturing
- Power generation and energy storage
- High-temperature electronics
- Radiation hardened electronics
- System-level thermal management M&S
- Thermal management components
- High-OPR compressors

PCA12: Processing-Enabled Intelligent ISR Sensors

- Chip-scale atomic clocks
- Virtual machine architectures
- Complex system dynamics
- V&V for complex adaptive systems
- Quantum computing
- Decision support tools
- Advanced RF apertures
- EO-IR sensing
- Distributed sensing networks
- Integrated sensing and processing
- Sensor-based processing
- Signal identification and recognition
- Information fusion and understanding

- Cyber offense
- Cyber defense
- Cyber resilience
- Advanced computing architectures
- Biological signatures
- Massive analytics
- Autonomous reasoning and learning
- Optical and infrared materials
- RF and electronic materials
- Metamaterials
- Nanomaterials
- High-temperature electronics
- Radiation hardened electronics

PCA13: High-Altitude, Long-Endurance ISR Airships

- Advanced aerodynamic configurations
- Aerodynamic T&E
- Cold-atom INSs
- Chip-scale atomic clocks
- Frequency-agile RF systems
- Spectral mutability
- DE protection
- Lightweight multifunctional structures
- Structural M&S
- Multiscale simulation technologies
- Coupled multiphysics simulations
- Validation support to simulations
- Autonomous systems
- Autonomous mission planning
- Embedded diagnostics
- Health monitoring and prognosis
- High-altitude airship

- Advanced RF apertures
- Secure RF links
- RF EW
- EO-IR sensing
- High-altitude materials
- Lightweight materials
- Advanced composites
- Metamaterials
- Nanomaterials
- Self-healing materials
- Material-specific manufacturing
- Power generation and energy storage
- High-temperature electronics
- Radiation hardened electronics
- System-level thermal management M&S
- Thermal management components

PCA14: Prompt Theater-Range ISR/Strike Systems

- Advanced aerodynamic configurations
- Aerodynamic T&E
- Cold-atom INSs
- Chip-scale atomic clocks
- Advanced thermal protection system (TPS) materials
- Scramjet propulsion systems
- Frequency-agile RF systems
- Spectral mutability
- DE protection
- Lightweight multifunctional structures
- Advanced composite fabrication
- Structural M&S
- Multiscale simulation technologies
- Coupled multiphysics simulations
- Validation support to simulations
- Autonomous systems
- Autonomous mission planning
- Embedded diagnostics
- Health monitoring and prognosis
- Advanced RF apertures
- Secure RF links
- EO-IR sensing
- IR signature suppression
- High-temperature materials
- High-altitude materials
- Lightweight materials
- Advanced composites
- Composites sustainment
- Material-specific manufacturing
- Hydrocarbon boost engine
- System-level thermal management M&S
- Thermal management components
- High-temperature fuel technologies
- Efficient bleedless inlets

PCA15: Fractionated, Survivable, Remotely Piloted Systems

- Advanced aerodynamic configurations
- Aerodynamic T&E
- Cold-atom INSs
- Chip-scale atomic clocks
- Ad hoc networks
- Polymorphic networks
- Complex adaptive distributed networks
- Laser communications
- Short-range secure RF communications
- Frequency-agile RF systems
- Complex adaptive systems
- Complex system dynamics
- V&V for complex adaptive systems
- DE effects
- DE protection
- High-power microwaves
- Lightweight multifunctional structures
- Advanced composite fabrication
- Structural M&S
- Multiscale simulation technologies
- Coupled multiphysics simulations
- Validation support to simulations
- Autonomous systems
- Autonomous reasoning and learning
- Resilient autonomy
- Collaborative/cooperative control
- Autonomous mission planning
- Embedded diagnostics
- Health monitoring and prognosis
- Advanced RF apertures
- Secure RF links
- RF EW
- EO-IR sensing
- IR signature suppression
- Cyber resilience
- Lightweight materials
- Advanced composites
- Composites sustainment
- Metamaterials
- Self-healing materials
- Material-specific manufacturing
- Power generation and energy storage
- System-level thermal management M&S
- Thermal management components
- Three-stream engine architectures
- High-OPR compressors
- Advanced/interturbine burners
- Efficient bleedless inlets
- Serpentine nozzles

PCA16: Direct Forward Air Delivery and Resupply

- Advanced aerodynamic configurations
- Aerodynamic T&E
- Cold-atom INSs
- Short-range secure RF communications
- Lightweight multifunctional structures
- Advanced composite fabrication
- Structural M&S
- Multiscale simulation technologies
- Coupled multiphysics simulations
- Validation support to simulations
- Autonomous systems
- Resilient autonomy
- Embedded diagnostics
- Health monitoring and prognosis
- Decision support tools
- Advanced RF apertures
- Secure RF links
- Lightweight materials
- Advanced composites
- Composites sustainment
- Material-specific manufacturing

PCA17: Energy-Efficient, Partially Buoyant Cargo Airlifters

- Advanced aerodynamic configurations
- Aerodynamic T&E
- DE effects
- DE protection
- Lightweight multifunctional structures
- Advanced composite fabrication
- Structural M&S
- Multiscale simulation technologies
- Coupled multiphysics simulations
- Validation support to simulations
- Autonomous systems
- Autonomous mission planning
- Embedded diagnostics
- Health monitoring and prognosis
- Advanced RF apertures
- Lightweight materials
- Advanced composites
- Composites sustainment
- Metamaterials
- Self-healing materials
- Nanomaterials
- Material-specific manufacturing
- Power generation and energy storage
- System-level thermal management M&S
- Thermal management components

PCA18: Fuel-Efficient Hybrid Wing-Body Aircraft

- Advanced aerodynamic configurations
- Aerodynamic T&E
- DE effects
- DE protection
- Lightweight multifunctional structures
- Advanced composite fabrication
- Structural M&S
- Multiscale simulation technologies
- Coupled multiphysics simulations
- Validation support to simulations
- Autonomous systems
- Embedded diagnostics
- Health monitoring and prognosis
- Advanced RF apertures
- IR signature suppression
- High-temperature materials
- Lightweight materials
- Advanced composites
- Composites sustainment
- Metamaterials
- Self-healing materials
- Nanomaterials
- Material-specific manufacturing
- System-level thermal management M&S
- Thermal management components
- Three-stream engine architectures

PCA19: Next-Generation High-Efficiency Turbine Engines

- Lightweight multifunctional structures
- Structural M&S
- Multiscale simulation technologies
- Coupled multiphysics simulations
- Validation support to simulations
- Embedded diagnostics
- Health monitoring and prognosis
- High-temperature materials
- Lightweight materials
- Nanomaterials
- Nondestructive evaluation
- Material-specific manufacturing
- High-temperature electronics
- Alternate fuels
- System-level thermal management M&S
- Thermal management components
- Three-stream engine architectures
- High-temperature fuel technologies
- High-OPR compressors
- Engine component testing
- Advanced/interturbine burners
- Serpentine nozzles

PCA20: Embedded Diagnostic/Prognostic Subsystems

- Cold-atom INSs
- Chip-scale atomic clocks
- Complex adaptive distributed networks
- Short-range secure RF communications
- Complex adaptive systems
- Complex system dynamics
- V&V for complex adaptive systems
- Fiber lasers
- Semiconductor lasers
- Lightweight multifunctional structures
- Multiscale simulation technologies
- Coupled multiphysics simulations
- Validation support to simulations
- Autonomous systems
- Autonomous reasoning and learning
- Resilient autonomy
- Embedded diagnostics
- Health monitoring and prognosis
- Decision support tools
- Advanced RF apertures
- Distributed sensing networks
- Integrated sensing and processing
- Signal identification and recognition
- Information fusion and understanding
- Cyber resilience
- High-temperature materials
- Lightweight materials
- Optical and infrared materials
- RF and electronic materials
- Metamaterials
- Nanomaterials
- Nondestructive evaluation
- Material-specific manufacturing
- Power generation and energy storage
- High-temperature electronics

PCA21: Penetrating, Persistent Long-Range Strike

- Advanced aerodynamic configurations
- Aerodynamic T&E
- Ad hoc networks
- Polymorphic networks
- Complex adaptive distributed networks
- Laser communications
- Short-range secure RF communications
- Frequency-agile RF systems
- Spectral mutability
- Complex adaptive systems
- Complex system dynamics
- V&V for complex adaptive systems
- DE effects
- DE protection
- Lightweight multifunctional structures
- Advanced composite fabrication
- Structural M&S
- Multiscale simulation technologies

- Coupled multiphysics simulations
- Validation support to simulations
- Autonomous systems
- Autonomous reasoning and learning
- Resilient autonomy
- Collaborative/cooperative control
- Autonomous mission planning
- Embedded diagnostics
- Health monitoring and prognosis
- Decision support tools
- Passive radar
- Advanced RF apertures
- Secure RF links
- RF EW
- EO-IR sensing
- IR signature suppression
- Integrated sensing and processing
- Information fusion and understanding
- Cyber offense

- Cyber resilience
- High-temperature materials
- Lightweight materials
- Advanced composites
- Composites sustainment
- Optical and infrared materials
- RF and electronic materials
- Metamaterials
- Self-healing materials
- Nanomaterials
- Material-specific manufacturing
- System-level thermal management M&S
- Thermal management components
- Three-stream engine architectures
- High-OPR compressors
- Advanced/interturbine burners
- Efficient bleedless inlets
- Serpentine nozzles

PCA22: High-Speed Penetrating Cruise Missile

- Advanced aerodynamic configurations
- Aerodynamic T&E
- Cold-atom INSs
- Chip-scale atomic clocks
- Advanced TPS materials
- Scramjet propulsion systems
- Frequency-agile RF systems
- Complex adaptive systems
- Complex system dynamics
- V&V for complex adaptive systems
- DE protection
- Lightweight multifunctional structures
- Advanced composite fabrication
- Structural M&S
- Multiscale simulation technologies
- Coupled multiphysics simulations
- Validation support to simulations
- Autonomous systems
- Autonomous mission planning
- Embedded diagnostics
- Health monitoring and prognosis

- Advanced RF apertures
- Secure RF links
- RF EW
- Integrated sensing and processing
- Signal identification and recognition
- Information fusion and understanding
- Cyber resilience
- High-temperature materials
- Lightweight materials
- Advanced composites
- Composites sustainment
- Metamaterials
- Self-healing materials
- Nanomaterials
- Material-specific manufacturing
- High-temperature electronics
- System-level thermal management M&S
- Thermal management components
- Engine component testing
- High-speed turbines
- Efficient bleedless inlets

PCA23: Hyperprecision Low-Collateral-Damage Munitions

- Advanced aerodynamic configurations
- Aerodynamic T&E
- Cold-atom INSs
- Chip-scale atomic clocks
- Ad hoc networks
- Short-range secure RF communications
- Complex adaptive systems
- V&V for complex adaptive systems
- Lightweight multifunctional structures
- Advanced composite fabrication
- Structural M&S
- Multiscale simulation technologies
- Coupled multiphysics simulations
- Validation support to simulations
- Autonomous systems
- Autonomous reasoning and learning
- Resilient autonomy
- Collaborative/cooperative control
- Autonomous mission planning
- Embedded diagnostics
- Advanced RF apertures
- Secure RF links
- EO-IR sensing
- Integrated sensing and processing
- Cyber resilience
- Lightweight materials
- Advanced composites
- Nanomaterials
- Material-specific manufacturing

PCA24: Directed Energy for Tactical Strike/Defense

- Advanced aerodynamic configurations
- Aerodynamic T&E
- Complex adaptive systems
- Complex system dynamics
- V&V for complex adaptive systems
- Solid-state lasers
- Fiber lasers
- Semiconductor lasers
- Beam control
- DE effects
- DE protection
- High-power microwaves
- Lightweight multifunctional structures
- Advanced composite fabrication
- Structural M&S
- Multiscale simulation technologies
- Coupled multiphysics simulations
- Validation support to simulations
- Embedded diagnostics
- Health monitoring and prognosis
- High-temperature materials
- Lightweight materials
- Advanced composites
- Power generation and energy storage
- High-temperature electronics
- System-level thermal management M&S
- Thermal management components
- Three-stream engine architectures

PCA25: Enhanced Underground Strike with Conventional Munitions

- Cold-atom INSs
- Chip-scale atomic clocks
- Structural M&S
- Multiscale simulation technologies
- Coupled multiphysics simulations
- Validation support to simulations
- Collaborative/cooperative control
- Autonomous mission planning

PCA26: Reusable Air-breathing Access-to-Space Launch

- Advanced aerodynamic configurations
- Aerodynamic T&E
- Advanced TPS materials
- Scramjet propulsion systems
- Complex adaptive systems
- Complex system dynamics
- V&V for complex adaptive systems
- Lightweight multifunctional structures

- Advanced composite fabrication
- Structural M&S
- Multiscale simulation technologies
- Coupled multiphysics simulations
- Validation support to simulations
- Embedded diagnostics
- Health monitoring and prognosis
- Advanced RF apertures
- Secure RF links
- Cyber resilience
- High-temperature materials
- High-altitude materials
- Lightweight materials
- Advanced composites
- Composites sustainment
- Self-healing materials
- Nanomaterials
- Material-specific manufacturing
- Hydrocarbon boost engine
- High-temperature electronics
- Radiation hardened electronics
- System-level thermal management M&S
- Thermal management components
- Efficient bleedless inlets

PCA27: Rapidly Composable Small Satellites

- Complex adaptive distributed networks
- Modular small-sat components
- Fractionated small-sat architectures
- Laser communications
- Short-range secure RF communications
- Frequency-agile RF systems
- DE effects
- DE protection
- Space weather
- Orbital environment characterization
- Satellite drag modeling
- SSA
- Lightweight multifunctional structures
- Advanced composite fabrication
- Structural M&S
- Multiscale simulation technologies
- Coupled multiphysics simulations
- Validation support to simulations
- Embedded diagnostics
- Health monitoring and prognosis
- Advanced RF apertures
- Secure RF links
- RF EW
- EO-IR sensing
- Integrated sensing and processing
- Cyber resilience
- High-altitude materials
- Lightweight materials
- Advanced composites
- Optical and infrared materials
- RF and electronic materials
- Metamaterials
- Nanomaterials
- Material-specific manufacturing
- Hydrocarbon boost engine
- Spacecraft propulsion
- Electric propulsion
- Power generation and energy storage
- High-temperature electronics
- Radiation hardened electronics
- System-level thermal management M&S
- Thermal management components

PCA28: Fractionated/Distributed Space Systems

- Ad hoc networks
- Polymorphic networks
- Agile networks
- Complex adaptive distributed networks
- Modular small-sat components
- Distributed small-sat architectures
- Fractionated small-sat architectures
- Laser communications
- Short-range secure RF communications
- Frequency-agile RF systems
- Spectral mutability
- Complex adaptive systems
- Complex system dynamics
- V&V for complex adaptive systems
- DE effects
- DE protection

- High-power microwaves
- Space weather
- Orbital environment characterization
- Satellite drag modeling
- SSA
- Lightweight multifunctional structures
- Advanced composite fabrication
- Structural M&S
- Multiscale simulation technologies
- Coupled multiphysics simulations
- Validation support to simulations
- Autonomous systems
- Collaborative/cooperative control
- Autonomous mission planning
- Embedded diagnostics
- Health monitoring and prognosis
- Advanced RF apertures
- Secure RF links
- RF EW
- EO-IR sensing
- Distributed sensing networks
- Integrated sensing and processing
- Cyber resilience
- High-temperature materials
- High-altitude materials
- Lightweight materials
- Advanced composites
- Optical and infrared materials
- RF and electronic materials
- Metamaterials
- Self-healing materials
- Nanomaterials
- Hydrocarbon boost engine
- Spacecraft propulsion
- Electric propulsion
- Power generation and energy storage
- High-temperature electronics
- Radiation hardened electronics
- System-level thermal management M&S
- Thermal management components

PCA29: Persistent Space Situational Awareness

- Distributed small-sat architectures
- Fractionated small-sat architectures
- Multiscale simulation technologies
- Coupled multiphysics simulations
- Validation support to simulations
- Decision support tools
- Advanced RF apertures
- EO-IR sensing
- Distributed sensing networks
- Integrated sensing and processing
- Information fusion and understanding
- Cyber resilience
- Complex environment visualization
- Massive analytics
- Optical and infrared materials
- RF and electronic materials

PCA30: Improved Orbital Conjunction Prediction

- Space weather
- Orbital environment characterization
- Satellite drag modeling
- SSA
- Multiscale simulation technologies
- Coupled multiphysics simulations
- Validation support to simulations
- Decision support tools
- Distributed sensing networks
- Information fusion and understanding
- Advanced computing architectures
- Complex environment visualization
- Massive analytics

Not every KTA listed under each PCA is necessarily relevant to every line of research that might support development of materiel or non-materiel means for achieving capabilities within that area. However, all are potentially relevant to at least some key aspects of approaches that might be pursued under that capability area.

Alignment of Key Technology Areas with Overarching Themes

Chapter 4 identifies a set of 12 broad overarching themes that helps to define the most essential research efforts that Air Force S&T should pursue to prepare for the strategic environment and enduring realities of the 2010–30 time frame. These overarching themes are aggregated into logical clusters below, and the KTAs identified above are mapped into these to clarify how these themes can be used to guide S&T investment choices.

FROM . . . Platforms, Integrated, Preplanned, Long System Life
TO . . . Capabilities, Fractionated, Composable, Expendable

- Advanced aerodynamic configurations
- Aerodynamic T&E
- Cold-atom INSs
- Chip-scale atomic clocks
- Ad hoc networks
- Polymorphic networks
- Virtual machine architectures
- Agile hypervisors
- Agile networks
- Pseudorandom network recomposition
- Complex adaptive distributed networks
- Modular small-sat components
- Distributed small-sat architectures
- Fractionated small-sat architectures
- Laser communications
- Short-range secure RF communications
- Frequency-agile RF systems
- Spectral mutability
- Dynamic spectrum access
- QKD
- Complex adaptive systems
- Complex system dynamics
- V&V for complex adaptive systems
- Solid-state lasers
- Fiber lasers
- Semiconductor lasers
- Beam control
- DE effects
- DE protection
- High-power microwaves
- Quantum computing
- Space weather
- Orbital environment characterization
- Satellite drag modeling
- SSA
- Lightweight multifunctional structures
- Advanced composite fabrication
- Structural M&S
- Multiscale simulation technologies
- Coupled multiphysics simulations
- Validation support to simulations
- Autonomous systems

- Autonomous reasoning
- Resilient autonomy
- Collaborative/cooperative control
- Autonomous mission planning
- Embedded diagnostics
- Health monitoring and prognosis
- Decision support tools
- Automated software generation
- Advanced RF apertures
- Secure RF links
- RF EW
- EO-IR sensing
- Distributed sensing networks
- Integrated sensing and processing
- Sensor-based processing
- Signal identification and recognition
- Information fusion and understanding
- Cyber resilience
- Advanced computing architectures
- Human behavior modeling
- Cultural behavior modeling
- Social network modeling
- Behavior prediction and anticipation
- Influence measures
- Cognitive modeling
- Complex environment visualization
- Massive analytics
- Automated reasoning and learning
- Cognitive performance augmentation
- Physical performance augmentation
- Human-machine interfaces
- Lightweight materials
- Advanced composites
- Composites sustainment
- Optical and infrared materials
- RF and electronic materials
- Metamaterials
- Self-healing materials
- Nanomaterials
- Nondestructive evaluation
- Material-specific manufacturing
- Spacecraft propulsion
- Electric propulsion
- Power generation and energy storage
- High-temperature electronics
- Radiation hardened electronics
- System-level thermal management M&S
- Thermal management components

FROM . . . Cyber Defense, Permissive, Operations, Single Domain TO . . . Cyber Resilience, Contested, Dissuasion, Cross Domain

- Cold-atom INSs
- Chip-scale atomic clocks
- Ad hoc networks
- Polymorphic networks
- Virtual machine architectures
- Agile hypervisors
- Agile networks
- Pseudorandom network recomposition
- Complex adaptive distributed networks
- Modular small-sat components
- Distributed small-sat architectures
- Fractionated small-sat architectures
- Laser communications
- Short-range secure RF communications
- Frequency-agile RF systems
- Spectral mutability
- Dynamic spectrum access
- QKD
- Complex adaptive systems
- Complex system dynamics
- V&V for complex adaptive systems
- DE effects
- DE protection
- High-power microwaves
- Quantum computing
- Space weather
- Orbital environment characterization
- Satellite drag modeling
- SSA
- Multiscale simulation technologies
- Coupled multiphysics simulations
- Validation support to simulations
- Autonomous systems
- Autonomous reasoning
- Resilient autonomy
- Collaborative/cooperative control
- Autonomous mission planning
- Embedded diagnostics

- Health monitoring and prognosis
- Decision support tools
- Automated software generation
- Passive radar
- Advanced RF apertures
- Secure RF links
- RF EW
- EO-IR sensing
- IR signature suppression
- Distributed sensing networks
- Integrated sensing and processing
- Sensor-based processing
- Signal identification and recognition
- Information fusion and understanding
- Cyber offense
- Cyber resilience
- Advanced computing architectures
- Biological signatures
- Human behavior modeling
- Cultural behavior modeling
- Social network modeling
- Behavior prediction and anticipation
- Influence measures

- Complex environment visualization
- Massive analytics
- Automated reasoning and learning
- Cognitive performance augmentation
- Physical performance augmentation
- Human-machine interfaces
- High-altitude materials
- Lightweight materials
- Advanced composites
- Composites sustainment
- Optical and infrared materials
- RF and electronic materials
- Metamaterials
- Self-healing materials
- Nanomaterials
- Spacecraft propulsion
- Electric propulsion
- Power generation and energy storage
- High-temperature electronics
- Radiation hardened electronics
- Alternate fuels
- System-level thermal management M&S
- Thermal management components

FROM . . . Sensors, Control, Single Domain
TO . . . Information, Autonomy, Cross Domain

- Cold-atom INSs
- Chip-scale atomic clocks
- Ad hoc networks
- Polymorphic networks
- Virtual machine architectures
- Agile hypervisors
- Agile networks
- Pseudorandom network recomposition
- Complex adaptive distributed networks
- Modular small-sat components
- Distributed small-sat architectures
- Fractionated small-sat architectures
- Laser communications
- Short-range secure RF communications
- Frequency-agile RF systems
- Spectral mutability
- Dynamic spectrum access
- QKD
- Complex adaptive systems
- Complex system dynamics

- V&V for complex adaptive systems
- DE protection
- High-power microwaves
- Quantum computing
- Space weather
- Orbital environment characterization
- Satellite drag modeling
- SSA
- Multiscale simulation technologies
- Coupled multiphysics simulations
- Validation support to simulations
- Autonomous systems
- Autonomous reasoning
- Resilient autonomy
- Collaborative/cooperative control
- Autonomous mission planning
- Embedded diagnostics
- Health monitoring and prognosis
- Decision support tools
- Automated software generation

- Secure RF links
- RF EW
- Distributed sensing networks
- Integrated sensing and processing
- Sensor-based processing
- Signal identification and recognition
- Information fusion and understanding
- Cyber resilience
- Advanced computing architectures
- Biological signatures
- Human behavior modeling
- Cultural behavior modeling
- Social network modeling
- Behavior prediction and anticipation

- Influence measures
- Cognitive modeling
- Complex environment visualization
- Massive analytics
- Automated reasoning and learning
- Cognitive performance augmentation
- Human-machine interfaces
- Metamaterials
- Self-healing materials
- Nanomaterials
- Power generation and energy storage
- High-temperature electronics
- Radiation hardened electronics

FROM ... . . . Manned, Control, Long System Life
TO . . . Remotely Piloted, Autonomy, Expendable

- Advanced aerodynamic configurations
- Aerodynamic T&E
- Cold-atom INSs
- Chip-scale atomic clocks
- Ad hoc networks
- Polymorphic networks
- Virtual machine architectures
- Agile hypervisors
- Agile networks
- Pseudorandom network recomposition
- Complex adaptive distributed networks
- Modular small-sat components
- Frequency-agile RF systems
- Spectral mutability
- Dynamic spectrum access
- Complex adaptive systems
- Complex system dynamics
- V&V for complex adaptive systems
- Semiconductor lasers
- DE effects
- DE protection
- High-power microwaves
- Quantum computing
- Lightweight multifunctional structures
- Advanced composite fabrication
- Structural M&S
- Multiscale simulation technologies
- Coupled multiphysics simulations
- Validation support to simulations

- Autonomous systems
- Autonomous reasoning
- Resilient autonomy
- Collaborative/cooperative control
- Autonomous mission planning
- Embedded diagnostics
- Health monitoring and prognosis
- Decision support tools
- Automated software generation
- High-altitude airship
- Passive radar
- Advanced RF apertures
- Secure RF links
- RF EW
- EO-IR sensing
- IR signature suppression
- Distributed sensing networks
- Integrated sensing and processing
- Sensor-based processing
- Signal identification and recognition
- Information fusion and understanding
- Cyber resilience
- Advanced computing architectures
- Human behavior modeling
- Cultural behavior modeling
- Social network modeling
- Behavior prediction and anticipation
- Cognitive modeling
- Complex environment visualization

- Massive analytics
- Automated reasoning and learning
- Human-machine interfaces
- High-altitude materials
- Lightweight materials
- Advanced composites
- Composites sustainment
- Optical and infrared materials
- RF and electronic materials
- Metamaterials
- Self-healing materials
- Nanomaterials
- Nondestructive evaluation

- Material-specific manufacturing
- Power generation and energy storage
- High-temperature electronics
- System-level thermal management M&S
- Thermal management components
- High-OPR compressors
- Engine component testing
- Advanced/interturbine burners
- Efficient bleedless inlets
- Serpentine nozzles
- High-speed turbines

FROM . . . Fixed, Single Domain, Permissive
TO . . . Agile, Cross Domain, Contested

- Advanced aerodynamic configurations
- Aerodynamic T&E
- Cold-atom INSs
- Chip-scale atomic clocks
- Ad hoc networks
- Polymorphic networks
- Virtual machine architectures
- Agile hypervisors
- Agile networks
- Pseudorandom network recomposition
- Complex adaptive distributed networks
- Distributed small-sat architectures
- Fractionated small-sat architectures
- Laser communications
- Short-range secure RF communications
- Frequency-agile RF systems
- Spectral mutability
- Dynamic spectrum access
- QKD
- Complex adaptive systems
- Complex system dynamics
- V&V for complex adaptive systems
- Solid-state lasers
- Fiber lasers
- Semiconductor lasers
- Beam control
- DE effects
- DE protection
- High-power microwaves
- Quantum computing

- Space weather
- Orbital environment characterization
- Satellite drag modeling
- SSA
- Lightweight multifunctional structures
- Multiscale simulation technologies
- Coupled multiphysics simulations
- Validation support to simulations
- Resilient autonomy
- Collaborative/cooperative control
- Autonomous mission planning
- Embedded diagnostics
- Health monitoring and prognosis
- Decision support tools
- Passive radar
- Advanced RF apertures
- Secure RF links
- RF EW
- EO-IR sensing
- IR signature suppression
- Distributed sensing networks
- Integrated sensing and processing
- Sensor-based processing
- Signal identification and recognition
- Information fusion and understanding
- Cyber offense
- Cyber defense
- Cyber resilience
- Advanced computing architectures
- Human behavior modeling

- Cultural behavior modeling
- Social network modeling
- Behavior prediction and anticipation
- Influence measures
- Cognitive modeling
- Complex environment visualization
- Massive analytics
- Automated reasoning and learning
- Cognitive performance augmentation
- Physical performance augmentation
- Human-machine interfaces
- Lightweight materials

- Optical and infrared materials
- RF and electronic materials
- Metamaterials
- Self-healing materials
- Nanomaterials
- Nondestructive evaluation
- Material-specific manufacturing
- Power generation and energy storage
- High-temperature electronics
- System-level thermal management M&S
- Thermal management components

Summary of Key Technology Areas

Among the 110 technology areas listed above, a smaller subset is common to many of these PCAs and overarching themes. This subset is especially important for enabling the shorter list of 12 top PCAs identified in the previous chapter. This set represents KTAs that are the most critical to advance over the next decade to enable Air Force capabilities that are matched to the 2030 environment.

- Autonomous systems
- Autonomous reasoning and learning
- Resilient autonomy
- Complex adaptive systems
- V&V for complex adaptive systems
- Collaborative/cooperative control
- Autonomous mission planning
- Cold-atom INSs
- Chip-scale atomic clocks
- Ad hoc networks
- Polymorphic networks
- Agile networks
- Laser communications
- Frequency-agile RF systems

- Spectral mutability
- Dynamic spectrum access
- QKD
- Multiscale simulation technologies
- Coupled multiphysics simulations
- Embedded diagnostics
- Decision support tools
- Automated software generation
- Sensor-based processing
- Behavior prediction and anticipation
- Cognitive modeling
- Cognitive performance augmentation
- Human-machine interfaces

The KTAs identified above represent only a fraction of the research efforts that the AFRL is currently pursuing. However, they are identified here as being among the most essential S&T that must be explored to enable Air Force capabilities that are matched to the 2010–30 environment. Other supporting research efforts beyond these will also be needed to obtain a properly balanced and hedged portfolio, but these areas will be essential. These KTAs will be needed to enable Air Force capabilities based on game-changing approaches such as flexible autonomy and human performance augmentation, which have been identified in *Technology Horizons* as central for maintaining Air Force technology superiority.

> *If we are to achieve results never before accomplished, we must expect to employ methods never before attempted.*
>
> —Sir Francis Bacon

Chapter 7

Grand Challenges for Air Force S&T
2010–30

This chapter presents four focused technology challenges to help guide S&T efforts and drive innovative solutions in ways that support the enduring realities, overarching themes, and KTAs identified in Technology Horizons. *Each leads to a demonstration of a substantial Air Force capability that requires significant S&T advances and integration across multiple technology areas.*

The following "grand challenges" are presented to provide a set of identifiable focus problems that can help guide Air Force S&T over the next decade in the directions outlined by *Technology Horizons*. These are candidate problems designed to promote integration of S&T efforts across traditional technical domains in ways that can trigger creative means to solve them. Each is intended to lead to a major systems-level integrated technology demonstration. The problem statements are written to be sufficiently descriptive of a major technology-enabled "stretch capability" without prescribing the approaches by which the challenge may be met.

These are not meant to be competitions in the way that some DOD-related grand challenges have been. They are focus areas within which sets of individual KTAs can be advanced and integrated at the whole-system level. The goal of each is to guide Air Force S&T in directions consistent with the enduring realities, overarching themes, and technology areas identified in *Technology Horizons*, as well as facilitate dramatic advances over current capabilities in areas that will be essential for the Air Force over the next two decades and beyond.

By design, the scale and scope of these challenges vary significantly. While the AFRL may be able to entirely solve some issues, others may require it to partner with other organizations. Each is intended to stress the solution space and drive innovative development of key technologies. Beyond advancing the underlying technologies and their integration via these focused efforts, the demonstrations will help broaden Air Force–wide awareness of S&T advances in ways that could potentially lead to accelerated development of fieldable capabilities derived from them.

Challenge #1:
Inherently Intrusion-Resilient Cyber Networks

The first challenge is to explore, develop, and demonstrate autonomous and scalable technologies that enable large, nonsecure networks to be made inherently and substantially more resilient to attacks entering through network or application layers and to attacks that pass through these layers. Emphasis is on advancing technologies that enable network-intrusion tolerance rather than traditional network defense, with the goal to achieve continued mission effectiveness under large-scale, diverse network attacks. Technologies may include, but are not limited to, virtualization, recomposability, IP hopping, pseudorandom switching, server self-cleansing, and other methods that provide substantial increases in resilience or tolerance to intrusions, are scalable to arbitrarily large networks, and do not require explicit network operator interventions. These technologies apply the overarching themes of agility, autonomy, composability, and resilience in cyber networks.

A representative synthesized, large-scale, open network should be used to demonstrate the individual or combined benefits of resulting technologies. This will include an extended period during which unrestricted parties worldwide are invited to achieve one or more of a set of defined goals that may include installation of malware or other implants, metatag alteration, data corruption, data exfiltration, distributed denial of service, and other representative predetermined objectives, in addition to provable accomplishment of any unspecified modifications in the network or interruptions of network functions.

Target date for demonstration: no later than 2015.

> *This challenge applies the overarching themes of agility, autonomy, composability, and resilience to cyber systems and will result in methods for achieving substantially improved intrusion resistance and intrusion tolerance in Air Force cyber systems.*

Challenge #2:
Trusted, Highly Autonomous
Decision-Making Systems

The second challenge is to explore, develop, and demonstrate technologies that enable current human-intensive functions to be replaced, in whole or in part, by more highly autonomous decision-making systems and technologies that permit reliable V&V to establish the needed trust in them. Emphasis is on decision-making systems requiring limited or no human intervention for current applications where substantial reductions in specialized manpower may be possible and for future applications involving inherent decision time scales far exceeding human capacity.

Technologies may include, but are not limited to, information fusion, cognitive architectures, robust statistical learning, search and optimization, automated reasoning, neural networks, complex system dynamics, and other approaches that will enable increasingly autonomous decision making. In parallel, generalized V&V methods should also be developed for highly autonomous, adaptive, near-infinite state systems that can provide asymptotic inferential assessments of confidence levels for such systems over their state space. Emphasis in both areas should be on identifying broadly applicable principles, supporting theoretical constructs, practical methods and their algorithmic embodiments, and measures of effectiveness, rather than focusing primarily on individual application-specific instantiations.

The effectiveness of the resulting technologies should be demonstrated in a representative synthesized, large-scale combined air operations center environment using recorded operational or operationally realistic data streams, with total or partial autonomous decision making for functions such as battle management, deployment planning, and logistics management. The resilience of decision-making systems should be verified under extremely high data-stream capacities and exceptional events, including corrupted data or information. Autonomous or semiautonomous decision quality should be quantified and compared with human operator decision-making results. Levels of trust in decision-making systems involved should also be quantified and compared with demonstration results.

Target date for demonstration: 2017.

> *Focused effort on this challenge will enable technologies that can support substantial manpower cost reductions and extend robust improved decision-making capabilities to highly stressing future applications that may involve decision time scales beyond human capacity.*

Challenge #3:
Fractionated, Composable, Survivable, Autonomous Systems

The third challenge is to explore, develop, and demonstrate technologies that can enable future autonomous aircraft or spacecraft systems achieving greater multirole capability across a broader range of missions at moderate cost, including increased survivability in contested environments. Emphasis is on composability via system architectures based on fractionation and redundancy. This involves advancing methods for collaborative control and adaptive autonomous mission planning, as well as V&V of highly adaptable, autonomous control systems. This also includes technology advances to enable autonomous coordinated flight operation of fractional elements using short-range, low-bandwidth, jam-resistant, secure communication links. This effort may include RF agility, burst transmissions, laser links, and other methods to achieve low probability of detection and maintain needed levels of link integrity. System-level optimization will determine the degree of fractionation and define functionality of each type of fractional element.

The demonstration should highlight underlying technologies in a fully integrated ground-based hardware-in-the-loop simulation of a complete fractionated system operating over a range of missions and denial environments. It should also incorporate representative fractional elements and level of redundancy needed for A2/AD operations. Active links between all system elements for data exchange and command and control communications should be featured. The demonstration should verify maintenance of link integrity over a range of jamming levels and broader EW attacks, as well as show resilience of system functionality to various types and degrees of degradation in fractional element capabilities. Finally, it should exhibit mission resilience to cyber attacks and to loss of multiple fractional elements.

Target date for demonstration: no later than 2018.

> *This challenge helps focus development of technologies to enable composable systems of RPAs from sets of common, relatively low-cost, fractional elements that cooperate as a single system to provide flexible capabilities across a broader range of missions.*

Challenge #4:
Hyperprecision Aerial Delivery
in Difficult Environments

The fourth challenge is to explore, develop, and demonstrate technologies that enable single-pass, extremely precise, autonomously guided aerial delivery of equipment and supplies under GPS-denied conditions from altitudes representative of operations in mountainous and contested environments and winds representative of steep, mountainous terrain. Emphasis is on low-cost, autonomous flight systems with control authority capable of reaching target point within specified impact limits under effects of large stochastic disturbances. Guidance technologies could include independent IMUs, back-referencing to delivery aircraft, or other methods. Technologies should address the full range needed to enable a complete air delivery system, including flight vehicle, guidance, terminal impact control, system health monitoring, and interfaces to airborne platform and ground units.

Resulting technologies should be demonstrated by their capability to consistently deliver payloads in excess of 2,000 pounds from a 25,000-foot altitude to within 25 meters of a designated point in steep terrain under wind conditions representative of mountainous areas. An integrated demonstration system should be deployable from current and future airlift systems, including remotely piloted platforms, in a single pass over the drop point. Impact metrics must meet standards for survivable delivery of equipment and supplies. The system should be linked to the Global Information Grid to report position, manifest, and system health information but remain resilient to cyber attacks. It must also be recoverable and, if possible, self-recoverable.

Target date for demonstration: 2018.

> *This challenge advances and integrates technologies in several key areas to demonstrate the potential of autonomous systems to meet the extreme requirements of a key Air Force mission.*

Chapter 8

Summary of *Technology Horizons* Vision

This chapter presents a summary of the S&T vision from Technology Horizons *and identifies the most critical focus areas for S&T investment during the next decade. Sustained research in these areas will provide the basis for Air Force capability dominance during 2010–30 and beyond.*

Technology Horizons has developed a realistically implementable vision of the most essential Air Force S&T that will be needed to meet the strategic, technological, and budget demands of 2010–30. The following recaps the range of inputs that went into developing this vision and its key elements and then describes the major insights from this effort.

Broad Range of Inputs to *Technology Horizons*

The most important S&T efforts that the Air Force will pursue over the next decade and beyond, as identified in this report, are the result of an exceptionally broad range of inputs that reflect

- inputs from working groups composed of individuals from the Air Force S&T and intelligence communities, MAJCOMs, product centers, FFRDCs, defense industry, and academia;

- broad S&T-related perspectives from visits, discussions, and briefings with organizations across the Air Force, including Air Staff and Air Force secretariat offices, MAJCOMs, product centers, direct reporting units, and field units;

- feedback obtained from visits, discussions, and briefings with organizations representing the DOD, federal agencies, FFRDCs, national laboratories, and industry;

- operational perspectives from briefings, visits, and discussions with ACC, AFSOC, AFSPC, and AMC; and

- technological, operational, and strategic perspectives in reports from the Air Force Scientific Advisory Board, Defense Science Board, Center for Strategic and Budgetary Assessments, and numerous other organizations.

Collectively, the broad range of perspectives that was considered in *Technology Horizons* ensures that the resulting S&T vision reflects key Air Force needs during 2010–30 as well as the principal technology-enabled opportunities where properly focused S&T investments can best meet those needs.

Elements of the S&T Vision

The role of any S&T vision is to help guide the right choices that must be made among a very large set of possible technology development efforts that could be pursued if resources were available. In the likely budget-constrained environment that the Air Force faces, such a guiding vision is essential for making these difficult choices. *Technology Horizons* thus provides an Air Force S&T vision for 2010–30 that consists of the following elements:

- *Strategic Context*: Principal strategic factors that will drive needed Air Force capabilities during this time and factors relevant to the worldwide S&T arena that will impact the Air Force's ability to maintain superior technological capabilities.

- *Enduring Realities*: Key drivers that will remain largely unchanged and that will act to constrain the ability of the Air Force to shape itself through S&T investments as it might otherwise be able to do in an unconstrained environment.

- *Overarching Themes*: Specific dominant themes that will be central for meeting Air Force needs resulting from the strategic context and enduring realities and that will form the essential foundation for the most important Air Force S&T during 2010–30.

- *Potential Capability Areas*: A set of technologically achievable capability areas that is well aligned with the overarching themes described above and that identifies key enabling technologies projected to be among the highest-value S&T investments during this time.

- *Key Technology Areas*: Cross-cutting technology areas that will be among the most essential for enabling key potential capabilities such as those above and that will be among the most important areas to emphasize in Air Force S&T over the next decade and beyond.

- *Grand Challenges*: A set of challenge problems to focus Air Force S&T over the next decade on many of the key enabling technology areas identified above, and to drive innovative approaches for solving the most important technical issues in these areas.

- *Implementation Plan and Recommendations*: The proposed plan for implementing the S&T vision from *Technology Horizons*, including actionable recommendations to vector Air Force S&T over the next decade, together with corresponding primary and supporting organizations to implement them.

Essential Focus Areas for Air Force S&T Investment

A major insight to emerge from *Technology Horizons* is that Air Force S&T over the next decade will need to focus as much on advancing technologies that enable reduced Air Force operating costs as on technologies that support more traditional development of new systems or capabilities. This includes technologies to reduce manpower, energy, and sustainment costs. Of these, manpower costs are by far the largest. Yet research specifically directed at increasing manpower efficiencies or reducing manpower needs has to date received substantially less attention as an identifiable Air Force S&T focus area.

While research into increased energy efficiency, improved energy storage, and alternative energy sources must continue, as must research to enable reduced sustainment costs, a significant new focus must be placed on technologies to achieve reduced manpower needs and greater manpower efficiencies. Regarding the latter, *Technology Horizons* has identified several areas in which substantial advances will be possible in the next decade through properly focused Air Force S&T investments: (1) autonomy and autonomous systems, (2) human performance augmentation, (3) freedom of operations in contested environments, and (4) intelligent sensors, laser and high-power microwave systems, persistent SSA, satellite systems, and gas turbine systems. All offer enormous potential for achieving capability increases and operating cost savings via improved manpower efficiencies and reduced manpower needs.

Increasing the Use of Autonomy and Autonomous Systems

A key finding is the need, opportunity, and potential to dramatically advance technologies that can allow the Air Force to gain capability increases, manpower efficiencies, and cost reductions through far greater use of autonomous systems in essentially all aspects of Air Force operations. Flexibly autonomous systems can be applied far beyond the current RPAs, operational flight programs, and other implementations in use today. Dramatically increased use of autonomy—not only in the number of systems and processes to which autonomous control and reasoning can be applied but especially in the degree of autonomy that is reflected in them—can offer the Air Force potentially enormous increases in its capabilities and, if implemented correctly, can do so in ways that enable manpower efficiencies and cost reductions.

Beyond the efficiency increases and associated manpower cost reductions that are achievable through such greater use of highly adaptible and flexibly autonomous systems and processes, significant time-domain operational advantages can be gained over adversaries who are limited to human planning and decision speeds. The increased ops tempo that is possible through greater use of autonomous systems itself represents a significant capability advantage.

Achieving these gains from greater use of autonomous systems will require development of entirely new methods for enabling "trust in autonomy" through V&V of the near-infinite state systems that result from high levels of adaptibility and autonomy. In effect, the number of possible input states that such systems can be presented with is so large that not only is it impossible to test all of them directly, but also it is not even possible to test more than an insignificantly small fraction of them. Proper development of such highly adaptive autonomous systems is thus inherently unverifiable by today's methods; as a result, their operation in all but comparatively trivial applications is uncertifiable.

Already today it is technically possible to imagine and build systems having immense levels of autonomy. But it is the lack of suitable V&V methods that prevents errors in the development of such systems from being detected, thereby keeping all but relatively low levels of autonomy from being certified for use. Potential adversaries, however, may be entirely willing to field systems that benefit from far higher levels of autonomy without imposing any need for certifiable V&V and could gain potentially significant capability advantages over the Air Force by doing so. The ease with which such highly autonomous systems can be designed and built in comparison with the substantial burden required to develop V&V methods for certifying them creates an inherent asymmetry that favors adversaries. Countering this asymmetric advantage will require access to as-yet-undeveloped methods for achieving certifiably reliable V&V.

Developing certifiable V&V methods for highly adaptive autonomous systems is one of the major challenges facing the entire field of control science and one that may require the larger part of a decade or more to develop a fundamental understanding of the underlying theoretical principles and various ways that they could be applied. The Air Force, as one the greatest potential beneficiaries of more highly adaptive and autonomous systems, must be a leader in the development of the underlying S&T principles for V&V.

Broadening the Augmentation of Human Performance

A second key finding to emerge from *Technology Horizons* is that natural human capacities are becoming increasingly mismatched to the enormous data volumes, processing capabilities, and decision speeds that technologies either offer or demand. Although humans today remain more capable than machines for many tasks, by 2030 machine capabilities will have increased to the point that humans will have become the weakest component in a wide array of systems and processes. Humans and machines will need to become far more closely coupled through improved human-machine interfaces and by direct augmentation of human performance.

The field of human performance augmentation has made considerable advances over the past two decades and—with appropriate focused research efforts over the next decade—will permit significant practical instantiations of augmented human performance. Such augmentation may come from increased use of autonomous systems as noted above, from improved man-machine interfaces to couple humans more closely and intuitively with automated systems, and even from direct augmentation of humans themselves. The latter includes neuropharmaceuticals or implants to improve memory, alertness, cognition, or visual/aural acuity, as well as screening of individuals for speciality codes based on brainwave patterns or genetic correlators, or even genetic modification itself. While such methods may be inherently distasteful to some, potential adversaries may be entirely willing to make use of them.

Developing acceptable ways of using S&T to augment human performance will become increasingly essential for gaining the benefits that many technologies can bring. The current technical maturity of various approaches in this area varies widely, but significant steps to advance and develop early implementations are possible over the next decade. Performance augmentation provides a further means to increase human efficiencies, allowing reduced manpower needs for the same capabilities or increased capabilities with given manpower.

Advancing Freedom of Operations in Contested Environments

A further key theme from *Technology Horizons* is the need to focus a greater fraction of Air Force S&T investments on research to support increased freedom of operations in contested or denied environments. Three main research areas are found to be of particular importance in this connection:

Cyber Resilience. Research into technologies for increased cyber resilience differs in subtle but important ways from the current emphasis being placed on cyber defense. While defense is focused on preventing adversaries from entering cyber systems, resilience refers to technologies that inherently make cyber systems far more difficult for an adversary to exploit once entry is gained, or that allow cyber systems to operate more effectively even when under overt cyber attack. Cyber resilience thus supports "fighting through" approaches that seek to maintain mission assurance across the entire spectrum of cyber threat levels.

Technologies such as massive virtualization, agile hypervisors, and inherent polymorphism can enable cyber systems that are fundamentally more resilient to intrusions. They greatly complicate a cyber adversary's ability to plan and coordinate effective cyber attacks by reducing the time over which our networks remain essentially static. In so doing they may also cause a cyber intruder to leave behind greater forensic evidence for attribution. It is likely that potential adversaries will also recognize the value of resilience in their own cyber networks, though our current greater dependence on cyber systems makes our development of such technologies more imperative.

The role of cyber operations will continue to grow as one of the foundational components of conflict escalation control and warfare. Beyond the defensive benefits of inherently resilient cyber environments, such technologies can also provide new means for expressing changes in defensive posture and signaling levels of escalation during periods of tension. For instance, progressively increasing the rate of network polymorphism as tensions increase permits the trading off of greater network performance to gain increased network resilience in a way that is readily detectable by an adversary and that allows signaling an increased state of preparedness to survive cyber attacks.

Precision Positioning, Navigation, and Timing in GPS-Denied Environments. The second focus area needed to enable increased freedom of operations is research on technologies to augment or supplant current precision PNT in GPS-denied environments. These include chip-scale IMUs and atomic clocks, as well as currently less mature "cold atom" INSs and timing systems based on compact matter-wave interferometry approaches.

The latter make use of various phenomena associated with Bose-Einstein condensates that can be realized with magneto-optical traps and other approaches currently being explored. The resulting ultraprecise position and timing capabilities may allow GPS-like accuracies relative to an initial reference to be maintained for sufficiently long periods to offset much of the benefit to an adversary of local GPS denial. In addition to such ultraprecise PNT augmentation, there is a need for research into improved terrain matching and other less accurate but robust approaches that can provide position information under broader GPS denial. The dependence of current Air Force systems on availability of PNT information makes research efforts to develop GPS surrogate technologies essential.

Electromagnetic Spectrum Warfare. The third research area to support freedom of operations in contested environments is in technologies for dominant EM spectrum warfare capabilities. These include various approaches for enabling greater spectral mutability to increase waveform diversity, including methods for pulse-to-pulse radar waveform encoding that can increase resilience to spoofing and resistance to signal injection. Dynamic spectrum access methods may also provide greater resilience to jamming and other modes of electronic attack and can give resilience to lost spectrum as bands are transferred to commercial uses. New methods for electronic attack are also needed to offset increasing adversary use of advanced integrated air defenses. Research on ultrawideband RF aperture technologies will be needed to allow spectrum warfare capabilities to be cost-effectively integrated into platforms.

Focusing on Additional High-Priority Technology Areas

Technology Horizons has further identified the following as being among key priority areas where S&T investment will be needed over the next decade to enable essential capabilities:

- *Processing-Enabled Intelligent Sensors*: ISR sensors with backplane processing for data synthesis and fusion to permit cueing-level PED functions to be performed on the sensor itself, reducing bandwidth otherwise consumed in transferring large amounts of raw data to the ground and manpower consumed for cueing-level processing.

- *Directed Energy for Tactical Strike/Defense*: Laser and high-power microwave systems with sufficient power and beam control for tactical strike, air vehicle defense, and active air base defense applications.

- *Persistent Space Situational Awareness*: Birth-to-death detection, tracking, and characterization of objects in orbit from traditional large satellites to picosatellites and orbital debris at LEO,

MEO, and GEO altitudes using ground- and space-based assets with data fusion from other sources into an integrated database.

- *Rapidly Composable Small Satellite Systems*: Satellites that can be assembled, tested, and launched within days of operational requirement, based on a plug-and-play/open-architecture approach using standards for self-describing components within a discoverable and autoconfiguring system.

- *Next-Generation High-Efficiency Gas Turbine Systems*: A wide range of technology advances that collectively can enable substantial improvements in turbine engine fuel efficiency needed for long-range strike and long-endurance ISR systems.

Several further key themes can be identified in the results from *Technology Horizons*. These include technology-derived means for achieving substantial fuel cost savings, in part through nontraditional flight vehicles such as hybrid wing-body aircraft, HALE airships, and partially buoyant cargo airlifters. While some such systems may not be fielded by the 2030 horizon date of this study, research to advance these technologies is appropriate to provide options when fuel prices become incompatible with the fuel efficiencies achievable by more conventional flight vehicles.

The future is here. It is just not widely distributed yet.

—William Gibson
Science fiction author who coined
the term "cyberspace" in 1984

Chapter 9

Implementation Plan and Recommendations

An implementation plan is presented with specific actionable recommendations for vectoring S&T in directions consistent with findings from Technology Horizons *to maximize the technology advantage of the US Air Force over the next decade and beyond.*

Recommendation #1:
Communicate Results from *Technology Horizons*

The first step involves communicating the rationale, objectives, and process for this study and the key elements of the resulting S&T vision across the Air Force in order to build broad awareness, understanding, and support for its implementation.

Brief *Technology Horizons* to Key Air Force Organizations

Technology Horizons should be offered as a brief to all HAF offices, MAJCOMs, and product centers, as well as to the AFRL and elsewhere, as appropriate, as a further means for disseminating its findings and seeking additional inputs.

Disseminate *Technology Horizons* Report

This report should be disseminated widely across key Air Force organizations beyond the HAF offices, MAJCOM and product center leadership, and Air Force S&T leadership that have provided inputs to the study. This effort should include operational and supporting components as a way of developing broad awareness, understanding, and support for Air Force S&T goals over the next decade.

Technology Horizons should also be disseminated across the DOD, including the director of defense research and engineering and DARPA; the other services, including the Office of Naval Research, Naval Research Laboratory, Army Research Office, and US Special Operations

Command; NASA and other federal agencies; as well as FFRDCs, academia, and industry organizations.

This may include briefings made to these organizations, industry groups, and others outlining the goals and S&T focus areas identified in *Technology Horizons*. These organizations will want to coordinate their S&T efforts with those of the Air Force and are likely to be partners in implementing some aspects of the Air Force S&T vision.

- Communicate *Technology Horizons* rationale, objectives, process, and key elements via briefings offered to all HAF offices, MAJCOMs, and product centers, as well as to the AFRL.

- Build broad awareness, understanding, and support for the Air Force S&T vision from *Technology Horizons*.

- Disseminate *Technology Horizons* across all relevant organizations beyond those that provided inputs to this effort.

> *Effectively communicating the Air Force S&T vision from* Technology Horizons *is the first step in developing broad awareness, understanding, support, and embracement of its objectives in order to enable its subsequent implementation steps.*

Recommendation #2:
Assess Alignment of S&T Portfolio with *Technology Horizons*

The second implementation step is for the AFRL to assess the alignment of its current S&T portfolio with the broad research directions and technology focus areas outlined in this report. This analysis would determine the extent to which the current portfolio is aligned and the fraction of the portfolio that should be aligned with these goals.

Determine Alignment of Current S&T Portfolio with *Technology Horizons*

Many of the technology focus areas where Air Force S&T should be focused to meet emerging Air Force needs are already being emphasized to various degrees in research that the AFRL is currently accomplishing. In other areas, insufficient or no work is being done on the S&T needed to enable these PCAs. Alignment of the current Air Force S&T portfolio with *Technology Horizons* should be assessed.

Identify Fraction of Portfolio to be Aligned with *Technology Horizons*

As noted throughout this report, no effort such as *Technology Horizons* can usefully list all S&T that is important for the Air Force to pursue. It must therefore be recognized that the KTAs identified here represent only a fraction of the overall research portfolio that the AFRL should pursue. In addition to assessing the alignment of its current S&T efforts with *Technology Horizons*, the Air Force should identify the fraction of its total S&T portfolio that it believes should be aligned with the focus areas identified here.

- Assess alignment of the AFRL's current S&T portfolio with the broad research directions and technology focus areas outlined in *Technology Horizons*.

- Identify the target fraction of the total Air Force S&T portfolio to be aligned with the research directions and technology focus areas identified in *Technology Horizons*.

> An appropriate fraction of Air Force S&T efforts should be consistent with key areas identified in Technology Horizons *as being essential to meet Air Force needs for 2010–30 and beyond.*

Recommendation #3: Adjust S&T Portfolio Balance As Needed

The third step is for the AFRL to highlight research efforts in its current S&T portfolio that require redirection or realignment with the vision in *Technology Horizons*, as well as new research that must be started to support these goals. These changes will redirect focus and emphasis in key areas.

Identify Current Efforts Requiring Realignment or Redirection

For the portion of its research portfolio that should be aligned with *Technology Horizons*, the AFRL will use its existing corporate process to identify current research efforts that are not well aligned with this vision. Furthermore, it will determine if an area can be redirected or realigned appropriately, or if it should be terminated to accommodate a more constructively formulated effort that achieves the needed direction and emphasis.

Determine New S&T Efforts That Must Be Started

Also using the existing corporate process, the AFRL will define new research efforts that enable the key goals identified in *Technology Horizons* to be effectively pursued. The resources needed to equip these pursuits may come from either the areas identified for realignment, redirection, or termination or from new resources dedicated specifically to enabling the vision in *Technology Horizons*.

- Identify research efforts in the current S&T portfolio that must be redirected or realigned with the research directions and technology focus areas in *Technology Horizons*.

- Determine which of these efforts should be realigned, redirected, or terminated to accommodate new research efforts that achieve the needed direction and emphasis.

- Define new research efforts that will be started to allow broad research directions and technology areas identified in *Technology Horizons* to be effectively achieved.

- Implement changes in the AFRL S&T portfolio to initiate new research efforts identified above, and to realign, redirect, and terminate existing efforts identified above.

> A mix of "buy, hold, and sell" positions will be needed among research efforts in the current Air Force S&T portfolio to achieve the S&T vision in Technology Horizons.

Recommendation #4:
Initiate Focused Research on
"Grand Challenge" Problems

As the next step after adjusting the current S&T portfolio, the AFRL should evaluate and refine a set of "grand challenge" problems similar to those identified here and use these to focus major technology development efforts in key areas identified in *Technology Horizons*. The specific challenge problems to be undertaken will be defined by the AFRL corporate process. However, they should be of a comparable scale and scope as those identified here, be structured to drive advances in key areas identified here as being essential for meeting emerging Air Force needs, and emphasize demonstrations that require sets of individual technology areas to be integrated at the whole-system level.

- Evaluate, define, and focus a set of grand-challenge problems of sufficient scale and scope to drive major technology development efforts in key areas identified here.

- Structure each grand challenge to drive advances in research directions identified in *Technology Horizons* as being essential for meeting Air Force needs in 2030.

- Define specific demonstration goals for each challenge that require sets of individual technology areas to be integrated and demonstrated at the whole-system level.

- Initiate sustained research efforts in the AFRL S&T portfolio as necessary to achieve each of the grand-challenge demonstration goals.

Properly choosing and sustaining a set of grand challenges is a key step in implementing the S&T vision identified in Technology Horizons *and for advancing a range of technologies that will be essential for meeting Air Force needs.*

Recommendation #5:
Improve Aspects of the Air Force
S&T Management Process

The final step involves improvements in two key aspects of the processes that the Air Force uses for planning and managing its S&T efforts.

Obtain HAF-Level Endorsement of a Stable
S&T Planning Construct

The "focused long-term challenges" process that the AFRL uses to manage its S&T portfolio was developed to enable effective near-, mid-, and long-range planning of research efforts. Some have suggested that a process organized around SCFs should replace this to better allow coordinating S&T efforts with broader Air Force objectives. Since transition to any new planning construct will entail substantial effort that will not directly contribute to research progress, such change should be done with deliberation and lead to HAF-level endorsement of the resulting planning construct. Air Force S&T will benefit significantly from the stability that such endorsement of an AFRL planning construct can bring, since this can provide the continuity of mid- and long-range planning needed to develop technologies that will be essential a decade or more from now.

Increase MAJCOM and Product Center Inputs
in the AFRL Planning Process

Although the planning construct used by the AFRL over the past five years involves regular inputs from all MAJCOMs and product centers, the level of involvement in providing such inputs has varied widely. For any AFRL planning construct to be effective in directing S&T to meet Air Force needs, it is essential that processes be established and consistently followed for obtaining such inputs. Regardless of the particular planning construct, this will require a formal process for obtaining periodic inputs from MAJCOMs and product centers to make adjustments at higher levels of S&T planning and programming, as well as a more frequent and less formal process at lower levels. The method for providing such inputs should be designed to ensure an appropriate balance between nearer-term responsiveness to the needs of MAJCOMs and product centers and longer-term responsiveness to S&T opportunities envisioned by the AFRL for meeting Air Force needs.

- Obtain HAF-level endorsement of an AFRL planning construct for S&T to provide the stability needed for effective mid- and long-range development of technologies.

- Define and implement a formal process for obtaining high-level inputs from MAJCOMs and product centers in periodic adjustments within the AFRL S&T planning construct.

- Develop and implement an informal process to obtain more frequent inputs from MAJCOMs and product centers for lower levels of the AFRL S&T planning construct.

These improvements to the S&T management process will ensure that research efforts over the next decade and beyond will properly account for MAJCOM and product center needs to maximize the technology advantage of the US Air Force.

Appendix A

Proposed Study for SECAF/CSAF Approval

USAF Chief Scientist Office

Proposed Study

Technology Horizons and Capability Implications for the Air Force

Background

The rapid "flattening" of the world from a technology perspective is allowing science and technology advances made anywhere to be exploited globally for developing militarily significant new capabilities. Many countries already have, or soon will have, the ability to translate worldwide technology advances into new offensive and defensive capabilities in the air, space, and cyber domains, and across domain boundaries. International markets in military systems will diffuse these capabilities rapidly and broadly. As a result, over the next two decades the U.S. will face a growing number of nations having near-peer or peer capabilities, and may find it increasingly difficult to maintain the technology superiority over potential adversaries that it has had in the past. Correctly anticipating those science and technology advances that will have greatest potential military significance—and the capabilities and counter-capabilities that may be derived from them—can help avoid technology surprise and ensure U.S. capability dominance.

This study will seek to identify key advances in science and technology that are likely to occur over the next 10 years that could in the following 10 years be developed into significant military capabilities. The use of this "10+10 technology-to-capability" forecasting process distinguishes this study from others in the Air Force and elsewhere that aim to understand various aspects of the opportunities and threats that emerging technologies present. Using this process, the study will develop a forward-looking yet realistic assessment on a 20-year horizon of potential offensive and defensive capabilities and counter-capabilities of the Air Force and its possible future adversaries.

Study Products

Briefing to SAF/OS & AF/CC in December 2009. Publish report in February 2010.

Charter

The study will:

- Conduct a "next-decade" (2020) assessment of technology advances that will be key to future air, space, and cyber domain capabilities, and to potential cross-domain capabilities.

- Provide a "following-decade" (2030) assessment of U.S. and adversary capabilities that could be developed from these technology advances, focusing on potential "leapfrog" and "game-changing" capabilities that may substantially alter future warfighting environments.

- Determine counter-capabilities that the Air Force will need in 2030 to be effective against these potential new adversary capabilities.

- Identify the underlying technologies that the Air Force will need in 2020 in order to develop the counter-capabilities it needs in 2030.

- Identify the science and technology research efforts that the Air Force must start today to develop the technologies it needs in 2020 to obtain the counter-capabilities it needs in 2030.

Appendix B

SECAF/CSAF Tasking Letter

THE SECRETARY OF THE AIR FORCE
CHIEF OF STAFF, UNITED STATES AIR FORCE
WASHINGTON DC

JUN 1 8 2009

MEMORANDUM FOR ALMAJCOM-FOA-DRU/CC
DISTRIBUTION C

SUBJECT: Technology Horizons Study

Air Force warfighting capabilities have a proud heritage of being born from the very best science and technology our Nation can create; indeed, the very history of the United States Air Force is closely intertwined with the development of advances in science and technology. Yet today, "flattening" of the world is making it increasingly challenging for the U.S. to maintain technology superiority over potential adversaries. A growing number of nations will soon have the ability to transform science and technology advances into militarily significant capabilities. Over the next decades, we will increasingly face potential adversaries having peer or near-peer capabilities. To remain the world's most capable Air Force, we must correctly anticipate the emerging science and technology advances that have the greatest military potential.

The Air Force Chief Scientist will conduct a study across the air, space, and cyberspace domains to develop a forward-looking assessment on a 20-year horizon of potential offensive and defensive capabilities and counter-capabilities of the Air Force and its future adversaries. This study will bring together scientists, engineers and operators from inside and outside the Air Force to develop a 10-year technology forecast, followed by a further 10-year forecast of new militarily significant capabilities that can be derived from those technologies. Using this "10+10 technology-to-capability" forecasting process, the study will seek to identify potential "leapfrog" and "game-changing" capabilities that could substantially alter future warfighting environments and affect future U.S. Joint capability dominance.

We believe this study can provide important insights in this pivotal time and we encourage support of its objectives. We expect that Air Force leaders at all levels may find the results useful in today's decision making as we work to ensure that our Air Force remains the world's most capable in 2030 and beyond.

Michael B. Donley
Secretary of the Air Force

Norton A. Schwartz
General, USAF
Chief of Staff

Appendix C

Chief Scientist of the Air Force
Transmittal Letter

DEPARTMENT OF THE AIR FORCE
HEADQUARTERS OF THE AIR FORCE
WASHNGTON DC

15 May 2010

The Honorable Michael B. Donley
Secretary of the Air Force
Air Force Pentagon
Washington, DC 20330

General Norton A. Schwartz
Chief of Staff of the U.S. Air Force
Air Force Pentagon
Washington, DC 20330

Re: "Technology Horizons": A Vision for Air Force Science & Technology

Secretary Donley, General Schwartz:

I am pleased to present the "Technology Horizons" final report for your consideration. This culminates a very substantial effort undertaken by the office of the Air Force Chief Scientist in accordance with your memorandum of 18 June 2009. "Technology Horizons" presents our vision of the key areas of science and technology that the Air Force must focus on over the next two decades to enable technologically achievable capabilities that can provide it with the greatest U.S. Joint force effectiveness by 2030.

The Air Force is at an undeniably pivotal time in its history, as the confluence of strategic changes, worldwide technological advancements, and looming resource constraints cause some to wonder how we will maintain our technological advantage. The "Technology Horizons" vision was crafted to help the Air Force vector its science and technology investments over the coming decade to focus more closely on addressing the complex strategic, technological, and budget challenges of 2010-2030.

The most essential insights from "Technology Horizons" regarding specific research focus areas may be summarized in approximate order of priority as follows:

1. During the coming decade, Air Force science and technology efforts will need to be focused as much on advancing technologies that can enable reduced Air Force operating costs as on technologies supporting more traditional development of new systems or capabilities.

2. These include technologies to reduce manpower, energy, and sustainment costs; of these, manpower costs are the largest, yet research specifically directed at increasing manpower efficiencies or reducing manpower needs has to date received substantially less attention as an identifiable Air Force focus area.

3. Two key areas in which significant advances are possible in the next decade with properly focused Air Force investment are: *(i)* increased use of autonomy and autonomous systems, and *(ii)* augmentation of human performance; both can achieve capability increases and cost savings via increased manpower efficiencies and reduced manpower needs.

4. Flexibly autonomous systems can be applied far beyond remotely-piloted aircraft, operational flight programs, and other implementations in use today; dramatically increased

use of autonomy -- not only in the number of systems and processes to which autonomous control and reasoning can be applied but especially in the degree of autonomy reflected in these -- offers potentially enormous increases in capabilities, and if implemented correctly can do so in ways that enable manpower efficiencies and cost reductions.

5. Greater use of highly adaptible and flexibly autonomous systems and processes can provide significant time-domain operational advantages over adversaries who are limited to human planning and decision speeds; the increased operational tempo that can be gained through greater use of autonomous systems itself represents a significant capability advantage.

6. Achieving these gains from use of autonomous systems will require developing new methods to establish "certifiable trust in autonomy" through verification and validation (V&V) of the near-infinite state systems that result from high levels of adaptibility; the lack of suitable V&V methods today prevents all but relatively low levels of autonomy from being certified for use.

7. The relative ease with which autonomous systems can be developed, in contrast to the burden of developing certifiable V&V methods, creates an asymmetric advantage to adversaries who may field such systems without any requirement for certifiability; countering this asymmetry will require access to as-yet undeveloped methods for establishing certifiably reliable V&V.

8. Developing V&V methods for highly adaptive autonomous systems is a major challenge facing the field of control science that may require a decade or more to solve; the Air Force, as one the greatest potential beneficiaries of such systems, must be a leader in developing the underlying science and technology principles for V&V.

9. Although humans today remain more capable than machines for many tasks, natural human capacities are becoming increasingly mismatched to the enormous data volumes, processing capabilities, and decision speeds that technologies offer or demand; closer human-machine coupling and augmentation of human performance will become possible and essential.

10. Augmentation may come from increased use of autonomous systems, interfaces for more intuitive and close coupling of humans and automated systems, and direct augmentation of humans via drugs or implants to improve memory, alertness, cognition, or visual/aural acuity, as well as screening for speciality codes based on brainwave patterns or genetic correlators.

11. Further key emphasis must be placed on research to support increased freedom of operations in contested or denied environments; three main research areas are found to be of particular importance in this connection: *(i)* cyber resilience, *(ii)* PNT in GPS-denied environments, and *(iii)* electromagnetic spectrum warfare.

12. While cyber defense seeks to prevent adversaries from entering cyber systems, resilience involves technologies that make cyber systems more difficult to exploit once entry is gained; cyber resilience supports "fighting through" to maintain mission assurance across the entire spectrum of cyber threat levels, including large-scale overt attacks.

13. Massive virtualization, agile hypervisors, and inherent polymorphism are technologies that can enable cyber systems to be fundamentally more resilient to intrusions; they complicate an adversary's ability to plan and coordinate attacks by reducing the time over which networks remain static, and cause an intruder to leave behind greater forensic evidence for attribution.

14. Beyond defensive benefits of inherently resilient cyber systems, the underlying technologies can also enable entirely new means in the cyber domain for expressing changes in defensive posture in ways that are intentionally detectable by an adversary to signal levels of escalation; such technologies offer new tools for cyber escalation control during periods of tension.

15. Research is needed on technologies to augment or supplant current precision navigation and timing (PNT) in GPS-denied environments; these include chip-scale inertial measurement units and atomic clocks, as well as currently less mature "cold atom" inertial navigation systems and timing systems based on compact matter-wave interferometry approaches.

16. Research is also needed into improved terrain matching and other less accurate but robust approaches that can provide position information under broader GPS denial; the dependence of current Air Force systems on availability of PNT information makes efforts to develop such GPS surrogate technologies essential.

17. Technologies to support dominant electromagnetic spectrum capabilities should include focus on methods for enabling greater spectral mutability to increase waveform diversity; critical among these are technologies for pulse-to-pulse radar waveform encoding that can increase resilience to spoofing and resistance to signal injection.

18. Development of dynamic spectrum access technologies in ways compatible with Air Force systems can give resilience to jamming and other modes of electronic attack, and provide flexibility needed when spectrum bands are lost to commercial uses; wideband RF aperture technologies will be needed to allow spectral mutability to be cost-effectively integrated.

19. Additional high-priority technology areas include: *(i)* processing-enabled intelligent sensors, *(ii)* directed energy for tactical strike/defense, *(iii)* persistent space situational awareness, *(iv)* rapidly composable small satellite systems, and *(v)* next-generation high-efficiency gas turbine engines; further technology areas to support fuel cost savings include hybrid wing-body aircraft, high-altitude long-endurance airships, and partially-buoyant cargo airlifters.

Beyond identifying these focus areas, "Technology Horizons" articulates a vision for Air Force science and technology that provides sufficient context and breadth to be a guiding document for the next decade and beyond. That vision consists of the following elements:

1. Strategic Context
2. Enduring Realities
3. Overarching Themes
4. Potential Capability Areas
5. Key Technology Areas
6. Grand Challenges
7. Vision Summary
8. Implementation Plan and Recommendations

While a properly hedged investment strategy is clearly called for, I believe that this vision will guide Air Force science and technology in areas that address the greatest challenges being faced during 2010-2030, and will help build broad awareness, understanding, and support throughout the Air Force for the increased technology focus that is needed to address these challenges.

Many individuals from a wide range of organizations provided inputs to "Technology Horizons." These spanned from MAJCOM and Air Staff level to operational squadron level, and included our sister Services, Department of Defense agencies, other Federal agencies, FFRDCs, national laboratories, industry and academia. This wealth of information and perspectives was distilled to identify the "disproportionately valuable" technology areas suited to the strategic, technological, and budget environments of 2010-2030. The results are presented in three volumes. Volume 1 presents the vision for Air Force S&T; two further volumes and an appendix provide additional supporting information.

In his 1945 "Towards New Horizons" report, Theodore von Kármán told General Hap Arnold that "only a constant inquisitive attitude toward science and a ceaseless and swift adaptation to new developments can maintain the security of this nation". That commitment to technological superiority has served the United States Air Force well over the intervening decades. Today, a vigorous and properly focused science and technology program remains absolutely essential to advancing the capabilities that the Air Force will need to fulfill its mission.

We must remain as committed as we were in 1945 to pursuing the most promising technological opportunities for our times, to having the scientific and engineering savvy to bring them to reality, and to having the wisdom to transition them into the next generation of capabilities that will allow us to maintain our edge. While we face substantial challenges in the coming decade, "Technology Horizons" has laid out a clear vision for the science and technology efforts that will be most essential for the Air Force. There is no greater organization than the United States Air Force to carry out this vision.

Very respectfully,

Werner J.A. Dahm
Chief Scientist of the U.S Air Force (AF/ST)
Air Force Pentagon

Attachments:
As stated

cc:
The Honorable E.C. Conaton (Under Secretary of the Air Force)
General C.H. Chandler (Vice Chief of Staff, U.S. Air Force)
LtGen W.C. Shelton (Assistant Vice Chief of Staff; U.S. Air Force)

Appendix D

SECAF/CSAF Cover Letter

MEMORANDUM FOR ALMAJCOM-FOA-DRU/CC   SEP 1 0 2010
DISTRIBUTION C

SUBJECT: Air Force Vision for Science and Technology

"Technology Horizons" outlines key science and technology focus areas for the next decade, and will be a valuable guide as we chart our path to the future. In this time of constrained budgets, the Air Force must determine which technologies offer the greatest return on potential investments. This perspective on the strategic, technological, and budgetary challenges that we face will serve as a useful baseline for determining the most promising technologies that will enable the Air Force to best operate in the air, space, and cyber domains.

The attached report presents a clearly articulated and grounded assessment of the strategic environment and enduring realities we face. It outlines a set of "overarching themes" that define system and technology attributes that are required to prevail. New technology-enabled capabilities are envisioned to meet key requirements, such as long-range strike, deterrence, cyber resilience, and energy efficiency, among others. The Air Force must move forward boldly to advance these technologies, through the dedicated, creative, and focused efforts of our science, technology, engineering, and mathematics workforce.

"Technology Horizons" will inform our Air Force Research Laboratory efforts on key technologies and their integration to realize militarily useful capabilities that meet Air Force and Joint needs. Working closely with our partners across the U.S. government, industry, academia, and partner nations, we will work to leverage the best intellectual capital and facilities in pursuing and advancing the most promising ideas. Accordingly, our science and technology strategies will posture the U.S. Air Force to sustain its technical and operational superiority.

We encourage all Airmen to read "Technology Horizons" to better understand these promising future technologies. The challenges that we face also present us with enormous opportunities. Together, we can ensure that technology can be harnessed effectively to continue providing our Nation with *Global Vigilance, Reach, and Power* in the 21st century.

Michael B. Donley
Secretary of the Air Force

Norton A. Schwartz
General, USAF
Chief of Staff

Appendix E

Working Groups

Appendix E

Working Groups

One source of inputs to *Technology Horizons* was a set of working groups specifically formed to provide a broad range of information, ideas, and viewpoints toward this effort.

Role of the Working Groups

Informal working groups were assembled for each of the air, space, and cyber domains, and an additional working group was assembled to address cross-domain topics. Each group served as a forum in which technical opinions of experts were provided to the Air Force chief scientist (AF/ST) and were discussed for possible further consideration in later phases of the effort, along with data and inputs from other sources.

Based on guidance from the Air Force general counsel (SAF/GC) to permit the free exchange of ideas and viewpoints that was essential to developing *Technology Horizons* while remaining consistent with federal rules that would otherwise apply to formal advisory committees, all working groups were formed on an ad hoc basis and were managed in a way that allowed sharing of individual opinions while specifically avoiding all efforts to reach any formal consensus among the group participants. Participants were involved in a strictly individual capacity; the opinions they expressed were their own views and did not necessarily represent the views of their organization. In addition, participants were specifically instructed not to discuss any proprietary contractor information. Discussions were of a broad, crosscutting nature, and no current contracts or specific Air Force requirements were discussed. There was no access to source selection information, contractor bid information, or proposal information. Operating the working groups in this manner gave the benefits of a broad range of views among the inputs to *Technology Horizons* without the rigid requirement of formal consensus that would have applied to formal advisory committees, which would have risked limiting the insights from these groups to largely predictable views.

Composition of the Working Groups

Participants for the working groups were selected in part for the combination of their scientific and technical background and knowledge, direct experience and knowledge of Air Force science and technology and Air Force operations, and ability to contribute effectively in a fast-paced, idea-intensive group environment. Participants were also selected to achieve reasonable coverage of the technical areas relevant to each domain. Each group was limited to roughly a dozen participants to allow for exchange of ideas and opinions. Working group participants were selected from the

- Air Force science and technology community,
- intelligence community,
- major commands,
- product centers,
- federally funded research and development centers,
- defense industry, and
- academia.

Since all participants strictly represented their own opinions and not their organizations' views, no attempt was made to achieve any specific target balance among types of organizations or to seek comprehensive representation of all organizations within any of these sectors.

The following describes the functions of each of the working groups and lists their contributors.

Air Domain Working Group

This working group considered technologies, concepts, and systems relevant to air domain functions, such as remotely piloted aircraft; fractionated strike system architectures; electronic warfare; bandwidth; GPS-denied environments; air domain threat systems; low-observable/counter-low-observable technologies; sensors; intelligence, surveillance, and reconnaissance systems; low collateral damage munitions;

air mobility; lighter-than-air and hybrid airships; base defense and force projection; power and thermal management; air-breathing hypersonic systems; near-space systems; and numerous other topics relevant to the air domain.

- Dr. Werner J. A. Dahm, Chair
- Col Eric Silkowski, PhD, Vice Chair
- Dr. Edward H. "Ned" Allen
- Douglas L. Bowers
- Dr. Gregory K. Crawford
- Dr. Donald R. Erbschloe
- Dr. Janet S. Fender
- Col Robert S. Fredell, PhD
- Dr. Mark A. Gallagher
- Dr. Thomas T. Hamilton
- Dr. Brian M. Kent
- Dr. Michael A. Kuliasha
- Betsy S. Witt
- Dr. John C. Zolper

Space Domain Working Group

This working group examined technologies and systems for space launch; communications; space system architectures; precision navigation and timing; space situational awareness; space protection; orbit transfer and maneuver; space-based intelligence, surveillance and reconnaissance; space radar; missile warning and intercept; space weather; space debris; small satellites; responsive space; laser communications; and other topics relevant to space domain functions.

- Dr. Werner J. A. Dahm, Chair
- Col Eric Silkowski, PhD, Vice Chair
- Dr. William L. Baker
- Dr. Alok Das
- Dr. Roberta Ewart
- Dr. Jonathan Gordon
- Dr. David A. Hardy
- Dr. Daniel Hastings
- Dr. Michael A. Kuliasha

- Gary A. O'Connell
- Dr. Jim F. Riker
- Dr. Lara S. Schmidt
- Dr. Dwight C. Streit

Cyber Domain Working Group

This working group examined offensive and defensive technologies and concepts relevant to the cyber domain, such as cyber situational awareness, cyber forensics and attribution, data protection, polymorphic networks, mobile ad hoc networks, information assurance, supervisory control and data acquisition systems, honeypots, deterrence, continuity of operations, mission assurance, cyber agility, attack containment, recovery, fight-through approaches, encryption, quantum key distribution, quantum computing, and other considerations relevant to the cyber domain.

- Dr. Werner J. A. Dahm, Chair
- Col Eric Silkowski, PhD, Vice Chair
- Dr. John S. Bay
- Richard J. Byrne
- Dr. Chris Colliver
- Jon Goding
- Dewey Houck
- Jeff A. Hughes
- Dr. Kamal T. Jabbour
- Dr. Michael A. Kuliasha
- Richard Mesic
- Dr. Marc Zissman

Cross-Domain Working Group

This working group considered technologies and systems that (1) have expected effects spanning across two or more of the air, space, and cyber domains, (2) have unexpected effects in a domain other than that in which expected effects occur, (3) may require key supporting functions from another domain, or (4) fall between these classical domains but have implications in one or more of these.

- Dr. Werner J. A. Dahm, Chair
- Col Eric Silkowski, PhD, Vice Chair
- Dr. Edward H. "Ned" Allen
- Dr. William L. Baker
- Richard J. Byrne
- Dr. Alok Das
- Dr. Richard P. Hallion
- Jeff A. Hughes
- Dr. Brian M. Kent
- Dr. Michael A. Kuliasha
- Dr. Jim F. Riker

Working Group Participant Biographies

The biographical summaries of the working group participants below give an overview of their respective expertise and background.

Dr. Edward H. "Ned" Allen

Dr. Allen is chief scientist of the Skunk Works at Lockheed Martin Aeronautics Company's Advanced Development Programs in Palmdale, CA, where he is also a corporate senior fellow. In his role as chief scientist, he represents the Skunk Works in various forums and serves as senior advisor for science and technology across the corporation. He is the principal inventor on more than two dozen patents and currently conducts research in areas that include energy, quantum information, hypersonic systems, hybrid airships, and micro air vehicles. After a decade on the tenured faculty of Utah State University and two years as a Rockefeller Foundation Fellow, he founded and ran his own commuter airline, now called Horizon Air, and then founded Daedalus Research Inc., an aeronautical research and development firm. In 1998 he brought his firm's key assets into Lockheed Martin's Skunk Works. He has degrees in physics from Stanford University, in mathematics and literature from Swarthmore College, and in applied mathematics and international relations from the University of Pennsylvania.

Dr. William L. Baker

Dr. Baker was chief scientist of the Directed Energy Directorate in the Air Force Research Laboratory, where as an internationally recognized leader in the entire spectrum of directed-energy technologies he has led research and development on high-energy laser weapons technology and the application of advanced optics to space situational awareness. He served as an Air Force nuclear research officer prior to becoming a civilian scientist at the Air Force Weapons Laboratory. He initially performed and led research on high-energy plasmas and pulse power systems for simulating nuclear weapon effects and then began work on high-energy particle beams for directed-energy weapon technologies. He led a joint effort to develop a unique accelerator that demonstrated stable beam propagation in open air. He then created and led the Air Force high-power microwave weapon technology program, which became the center of excellence for the nation and has demonstrated multiple military applications. He is a fellow of the Institute of Electrical and Electronics Engineers and president of the Directed Energy Professional Society. He has received the DOD Distinguished Civilian Service Award and the Distinguished Presidential Rank Award. He has BS and MS degrees in physics from Ohio State University and a PhD in nuclear physics from Ohio State University.

Dr. John S. Bay

Dr. Bay served as chief scientist of the Information Directorate in the Air Force Research Laboratory (AFRL), directing the planning and technical execution of information systems science and technology and its transition to air, space, and cyberspace systems. Prior to joining AFRL he was a program manager in the Information Exploitation Office (IXO) at the Defense Advanced Research Projects Agency, an engineering fellow at Raytheon Company, and a professor of electrical and computer engineering at Virginia Tech. His research focus is on control systems, robotics, biomechanics, machine learning, and embedded systems. He is an author of more than 65 publications, a fellow of the Institute of Electrical and Electronics Engineers (IEEE), and a winner of the 2009 IEEE Computer Society Technical Achievement Award for his work in embedded systems. He is a former IEEE Computer Society

Distinguished Visitor and is a 2005 recipient of the Office of the Secretary of Defense Medal for Exceptional Public Service. He holds a BS degree in electrical engineering from Virginia Tech and MS and PhD degrees from Ohio State University.

Douglas L. Bowers

Mr. Bowers is director of the Propulsion Directorate in the Air Force Research Laboratory (AFRL), where he oversees the Air Force's science and technology program in propulsion and power for space, missile, and aircraft applications. He previously was associate director for air platforms in AFRL's Air Vehicles Directorate, responsible for fixed-wing and rotary-wing technologies, hypersonics, turbine engines, and power. He has served in a variety of senior technical positions within the Air Force, leading the development and fielding of advanced highly survivable engine inlets and exhaust nozzles. He was a technical consultant on the C-17, F-15E, F-16, and B-1 development programs. He has served as a member of and as US national coordinator for the NATO Advisory Group for Aerospace Research and Development's fluid dynamics panel. As the sole Air Force staff member for the Commission on the Future of the US Aerospace Industry, he developed recommendations for aerospace industry health and national security. He has a BS degree in aerospace engineering from Purdue University, an MS degree in aerospace engineering from Ohio State University, and an MS degree in engineering management from the University of Dayton.

Richard J. Byrne

Mr. Byrne is a vice president at the MITRE Corporation with nearly 30 years of experience developing computing, communications, networks, information technologies, and systems engineering products and prototypes. His responsibilities include technical centers, research programs, and other engineering activities in the command and control center, as well as exploring solutions to national security problems with emphasis on improved information interoperability, systems integration, and cyber assurance, including net-centric strategies, complex systems engineering, and information technologies. He has held various positions at MITRE supporting the Air Force and DOD, including vice president for all Air Force programs and, previously, technical director for the Air Force Electronic Systems Center and executive director of innovation. He was a founder and engineering manager at a semiconductor startup, leading design and production release of more than 150 products over five years. He was technical manager of very-large-scale integration design methods for telecommunication at ITT's Advanced Technology Center. He has a BS in electrical engineering and an MS in electrical engineering and computer science from the Massachusetts Institute of Technology.

Dr. Chris Colliver

Dr. Colliver is the technical advisor for command, control, communications, and computer systems/information operations (C4/IO) analysis at the National Air and Space Intelligence Center (NASIC), where he leads intelligence analysis and production of foreign C4 intelligence, surveillance, and reconnaissance and cyberspace capabilities to support joint operations, acquisition, and policy making. He has worked in a number of analytic, advisory, and supervisory positions responsible for foreign space and counterspace systems analysis. He established a Counterspace Analysis Division and expanded NASIC's defense intelligence space order of battle to help lead the transformation of NASIC into a more space-centric organization. He served as the space topic manager for the National Intelligence Priorities Framework from 2004–8 and was named an exceptional intelligence analyst by the director of Central Intelligence. He has experience in analog and digital communications via high frequency, microwave, and satellite, as well as in network operations. He holds BS, MS, and PhD degrees, all in electrical engineering from the University of Dayton.

Dr. Gregory K. Crawford

Dr. Crawford is the director for integration of intelligence, surveillance, and reconnaissance (ISR) at MITRE, focused on ISR sensing, sensor data and signal processing, exploitation algorithm development and prototyping, and enterprise systems engineering. He also leads work on integrated sensing, processing, and exploitation as part of the MITRE corporate challenge, in which he guides 30 research and development (R&D) projects focused on operationally motivated capabilities directed at difficult problems within this domain. He has led multiple projects, with contributions ranging from basic R&D to system design and end-to-end systems engineering. Recently, he served as project leader and chief engineer for space radar battle management command and control, addressing a wide variety of ISR exploitation and enterprise-level systems engineering issues seeking to develop "dual-use" (tactical-national) capabilities. He is an author of more than 20 papers on a broad array of geophysical topics in leading journals and has written numerous technical papers on applied topics. He has a BS in physics and BA in chemistry from California State University, Fresno, an MS in geophysics and space physics from the University of California, Los Angeles (UCLA), and a PhD in space plasma physics from UCLA.

Dr. Werner J. A. Dahm

Dr. Dahm is the chief scientist of the US Air Force. He is the principal advisor for science and technology to the Air Force chief of staff and the secretary of the Air Force. He is on leave from the University of Michigan, where for the past 25 years he has served as a professor of aerospace engineering. His work has principally been in areas related to aerodynamics and propulsion. He is an author of over 180 journal articles, conference papers, and technical publications; a holder of several patents; and a prolific lecturer who has given over 120 invited, plenary, and keynote presentations worldwide on aerospace engineering topics. He has been a member of the Air Force Scientific Advisory Board and has served on numerous task forces for the Defense Science Board and as a member of the Defense Science Study Group. He is a fellow of both the American Physical Society and the American Institute of Aeronautics and Astronautics and a recipient of the William F. Ballhaus Aeronautics Prize from Caltech, as well as major research awards from the University of Michigan. He has served extensively in advisory and organizational roles in his field and as a consultant for industry. He has BS and MS degrees in mechanical engineering from the University of Alabama and University of Tennessee Space Institute and a PhD degree in aeronautics from Caltech.

Dr. Alok Das

Dr. Das is senior scientist for design innovation at the Air Force Research Laboratory (AFRL). As chief innovation officer at AFRL, he is the principal advisor for technology and process innovation strategies and leads the rapid reaction team that utilizes innovation and collaboration to develop near-term solutions to urgent war-fighter needs. He previously was chief scientist in AFRL's Space Vehicles Directorate, focusing on revolutionary space mission architectures, satellite designs, and technologies to reduce cost and increase capability and operating flexibility of space systems. He first was an engineer in the Indian Space Research Organization and then joined the Air Force Rocket Propulsion Laboratory to become technical lead for large space structures. He was instrumental in developing and transitioning smart structures technology from conceptual demonstrations to applications in space systems, giving satellite designers an array of space-proven technologies. He is a fellow of the American Institute of Aeronautics and Astronautics and an author of more than 60 technical articles on space technologies. He has a BS degree in electronics and communications engineering, an MS in aeronautical engineering from the Indian Institute of Science in Bangalore, and a PhD in aerospace engineering from Virginia Tech.

Dr. Donald R. Erbschloe

Dr. Erbschloe is the chief scientist for Air Mobility Command (AMC), where he advises the AMC commander and key leadership on scientific issues and technological enhancements of AMC capabilities. He has had a 28-year military career in the Air Force balanced among three primary thrusts: operations, academia, and scientific and technical management. He was a command pilot with 3,900 flying hours in the C-141, the TG-7A, and the UV-18. He has served as an associate professor of physics at the US Air Force Academy, chief scientist of the European Office of Aerospace Research and Development in the Air Force Office of Scientific Research, and director of faculty research at the Air Force Academy. He later served as military assistant to the chief scientist of the Air Force, and most recently he was acting chief operations officer for the Office of Science in the Department of Energy, which manages 10 world-class laboratories and is the single largest supporter of basic research in the physical sciences in the United States. He has a BA degree in physics and mathematics from the University of Virginia, an MS degree in physics from the Air Force Institute of Technology, and a PhD degree in physical electronics from Oxford University in England.

Dr. Roberta M. Ewart

Dr. Ewart is the chief scientist for the Air Force Space and Missile Systems Center in El Segundo, California. Her responsibilities include the identification, maturation, and transition of key technologies for national security space missions. Her career experience spans space flight operations with NASA and the USAF, as well as research, development, and acquisitions positions. Her research focus areas include solid-state lasers, photonics, laser communication, and dynamic radiation response modeling of spacecraft. She was named the 2008 Air Force Space Command senior civilian manager of the year. She holds a BS degree in physics from the United States Air Force Academy, degrees in theoretical physics and philosophy from Oxford University, and a degree in electrical engineering from Stanford University.

Dr. Janet S. Fender

Dr. Fender is the chief scientist in Air Combat Command (ACC), serving as scientific adviser to the commander and providing scientific expertise and technical guidance throughout ACC. She serves as the primary interface for ACC to the scientific community to identify technologies and catalyze enhancements to ACC's war-fighting capabilities. Her career in science and technology includes basic laboratory research, fielding new capabilities, key technical positions in major programs, and support to operations. Recently she led Air Force studies on "Day without Space" and the "Value of Stealth." She previously was chief scientist for the Space Vehicles Directorate in the Air Force Research Laboratory and senior scientist for directed energy and advanced imaging. She has authored over 100 technical papers, is a fellow of the Optical Society of America and the International Society of Optical Engineers, and serves on the executive committee of the board of directors of the American Institute of Physics. She has a BS degree in physics and astronomy from the University of Oklahoma and MS and PhD degrees in optical sciences from the University of Arizona.

Col Robert S. Fredell, PhD

Colonel Fredell is chief scientist and director of research at the US Air Force Academy, where he directs 13 scientific and engineering research centers and two institutes in the largest undergraduate university research program in the United States. He has performed basic and applied research in aircraft structural fatigue and composite structural repair and has taught engineering mechanics at the academy. He has expertise in advanced metallic and composite materials, with particular focus on fiber metal laminates and their application to new and aging aircraft. He previously served as the military assistant to the Air Force chief scientist in the Pentagon, deputy director of engineering at the Warner Robbins Air Logistics Center, director of faculty research at the US Air Force Academy, and technical director of the European Office of Aerospace Research and Development of the Air Force Office of Scientific Research. He has BS and MS degrees in mechanical engineering from Oklahoma State University and a PhD degree in aerospace engineering from Delft University in the Netherlands.

Dr. Mark A. Gallagher

Dr. Gallagher is technical director in Studies and Analyses, Assessments, and Lessons Learned (AF/A9) in Headquarters Air Force. In this role he supports analyses across the spectrum of Air Force missions and develops and applies modeling and simulation to support Air Force war-fighting and force structure capability and sufficiency assessments. He previously was a member of the full-time faculty at the Air Force Institute of Technology (AFIT), where he continues to serve as an adjunct associate professor. He has had prior assignments in the Cost Analysis Improvement Group of Program, Analysis, and Evaluation in the Office of the Secretary of Defense and as chief of the Missile Defense and Combating Weapons of Mass Destruction Capabilities Division at US Strategic Command. He founded the Information Operations *Joint Munitions Effectiveness Manual*, which he also chaired for four years. He is serving a second term on the Military Operations Research Society board of directors as vice president of society services. He holds BS degrees in operations research and computer science from the US Air Force Academy and MS and PhD degrees in operations research from AFIT.

Jon Goding

Mr. Goding is a principal engineering fellow with Raytheon's Network Centric Systems business. He presently serves as chief engineer for Raytheon's cyber initiative, where he is responsible for coordinating research and development in computer network operations technology across the company. In 23 years at Raytheon, he has designed computing and network systems for many DOD and federal government customers and was the architect for information assurance on the Navy / Marine Corps Internet from pre-award through initial operations. At the time it went operational, this was the largest integrated secure network in use. When Raytheon formed a new Secure Networks product line in 2003, he was named its technology director. During recent years, he has participated in many government and industry working groups and sponsored and advised numerous university research projects. He has a BS degree in electrical engineering from the University of Florida and an MS degree in electrical engineering from the University of South Florida.

Jonathan D. Gordon

Mr. Gordon is a program area chief engineer for the Advanced Concepts and Technology group at Raytheon Space and Airborne Systems (SAS), with engineering oversight for all the company's radio-frequency (RF) technology development programs and internal research and development efforts in this business area. He is also responsible for the development of long-range and "white space" technology roadmaps that guide internal investment and program pursuits. He has 30 years of technical and programmatic leadership at Raytheon and Hughes Aircraft Company. His primary expertise is active array subsystem design, integration, and test for both space and airborne applications. Prior to his chief engineer assignment, he was the manager of advanced

RF space programs in SAS, where he was involved in numerous space RF programs, including the European Space Agency front-end processor technology development program (space-based radar), innovative space-based radar antenna technology, Technical Satellite of the 21st Century (TechSAT 21), and Discoverer II, and has served as a technical and integrated product team lead for several active array subsystem engineering efforts. He has BS degrees in physics and in electrical engineering from the University of Southern California (USC) and an MS degree in electrical engineering from USC.

Dr. Richard P. Hallion

Dr. Hallion is an internationally recognized aerospace historian and founder of Hallion Associates, which evaluates emerging technologies and assesses their potential for application to meet civil and military aerospace needs. He retired from the United States Air Force in 2006 as senior advisor for air and space issues, served also as the Air Force historian for over a decade, and held the Harold Keith Johnson Visiting Chair in Military History at the US Army Military History Institute, Army War College. He was a founding curator of the National Air and Space Museum and subsequently held the Charles A. Lindbergh Chair in Aerospace History and an Alfred Verville Fellowship in aeronautics. He has written over a dozen works on the history of flight and military and civil aviation, teaches and lectures widely, and is a fellow of the Earth Shine Institute and a fellow of the American Institute of Aeronautics and Astronautics. He holds a BA degree and a PhD degree, both from the University of Maryland.

Dr. Thomas T. Hamilton

Dr. Hamilton is senior physical scientist in national security at the RAND Corporation, where since 1996 he has worked on a variety of issues for the DOD, with a focus on unmanned aircraft, airport security, aircraft test facilities, and force modernization. In particular, his work has addressed the future of USAF technology, aircraft survivability issues, unmanned aircraft system procurement and operations, USAF tanker modernization, force modernization and employment, aircraft test facilities, and airport security. Prior to joining RAND, from

1986 to 1996 he was a research astrophysicist at Harvard University, Columbia University, and the California Institute of Technology, and from 1981 to 1983, he developed satellite guidance software at Martin Marietta Corporation. He also served on the faculty in the Department of Astronomy at Caltech. He has an AB degree in physics from Harvard University and a PhD in physics from Columbia University.

Dr. David A. Hardy

Dr. Hardy is associate director for space technology in the Space Vehicles Directorate of the Air Force Research Laboratory (AFRL). He is responsible for oversight of space science and technology investments and for the senior civilian management of the Space Vehicles Directorate's science and technology portfolio. He leads AFRL's efforts across the full range of existing and emerging space technologies needed to support present and future space systems for the Air Force and DOD, including science and technology related to sensors, materials, information technology, space vehicles, directed energy, propulsion, and human effectiveness. He also serves as an adviser on space science and technology to Air Force Space Command, the Space and Missile Systems Center, the Defense Advanced Research Projects Agency, and other organizations and develops and maintains coordination in the growth and execution of space science and technology by national and international governmental agencies and academia. He has a BS degree in physics from Duke University and MS and PhD degrees in space physics and astronomy from Rice University.

Dr. Daniel Hastings

Dr. Hastings is a professor of aeronautics and astronautics at the Massachusetts Institute of Technology (MIT) and serves as MIT's dean for undergraduate education. He has taught courses in plasma physics, rocket propulsion, space power and propulsion, aerospace policy, and space systems engineering. He was the chief scientist of the US Air Force from 1997 to 1999, serving as the chief scientific adviser to the Air Force chief of staff and secretary of the Air Force. He led several influential studies on where the Air Force should invest in space, global energy projection, and options for a science and technology workforce for the twenty-first century. His research is on space systems and space policy and has also focused on issues related to spacecraft environmental interactions, space propulsion, space systems engineering, and space policy. He is a fellow of the American Institute of Aeronautics and Astronautics, a fellow of the International Council on Systems Engineering, a member of the International Academy of Astronautics, a past member of the National Science Board, and a former chair of the Air Force Scientific Advisory Board. He has a BA in mathematics from Oxford University in England and SM and PhD degrees in aeronautics and astronautics from MIT.

Dewey R. Houck II

Mr. Houck is a senior technical fellow working in Boeing's Integrated Defense Systems for the Mission Systems organization and, since July 2008, has been the director of the Mission Systems organization. He previously served as the chief technical officer for Mission Systems with primary responsibilities in the areas of technology strategy development, research planning, and integration of Mission Systems intellectual property into other Boeing programs. Mission Systems is the Boeing business unit that serves intelligence community customers including the National Security Agency, Central Intelligence Agency, National Geospatial-Intelligence Agency, and director of National Intelligence, with its focus on developing and deploying information management systems to facilitate tasking, collection, sharing, and analysis of intelligence. He previously served on the senior management team of Autometric, Inc. as vice president for technology development, where he administered all research and product development activities, including several geospatial, photogrammetric, and visualization initiatives. He has BS and MS degrees in civil engineering with specialization in photogrammetry and geodesy from Virginia Tech.

Jeff A. Hughes

Mr. Hughes is chief of the Autonomic Trusted Sensing for Persistent Intelligence (ATSPI) Technology Office in the Sensors Directorate of the Air Force Research Laboratory, where he leads research on complex system decomposition, vulnerability, and risk analysis to develop trusted systems. He was selected to establish the ATSPI Technology Office to support the Office of the Secretary of Defense software protection initiative to prevent unauthorized distribution and exploitation of application software and associated data critical to national security. ATSPI is the field office for the DOD Anti-Tamper Executive Agent, with the mission to deter the reverse engineering and exploitation of the US military's critical technologies. His prior work has included developing and deploying combat identification capabilities and low observable / counter-low observable radio-frequency and electronic warfare technology to Air Force weapons systems. He has served on numerous NATO and US government interagency committees and DOD working groups and has received numerous awards for his work. He has BS and MS degrees in electrical engineering from Ohio State University and has completed work toward his PhD at the Air Force Institute of Technology.

Dr. Kamal T. Jabbour

Dr. Jabbour is senior scientist for information assurance in the Information Directorate of the Air Force Research Laboratory (AFRL), where he serves as principal scientific advisor for information assurance, including defensive and offensive information warfare and technology. He plans and advocates major research and development activities, guides the quality of scientific and technical resources, and provides expert technical consultation to organizations across the Air Force, the DOD, government agencies, universities, and industry. He is program director of the Advanced Course in Engineering Cyber Security Boot Camp for developing future Air Force cyber warfare officers and leaders. He began his professional career on the computer engineering faculty at Syracuse University, where he taught and conducted research for two decades, including a three-year term as department chairman. In 1999 he joined the Cyber Operations Branch at AFRL through the Intergovernmental Personnel Act and transitioned gradually from academia to his current position. He has a BS degree in electrical engineering from the American University of Beirut in Lebanon and a PhD degree in electrical engineering from the University of Salford in England.

Dr. Brian M. Kent

Dr. Kent is chief scientist of the Sensors Directorate in the Air Force Research Laboratory, where he serves as principal scientific and technical adviser with primary authority for the technical content of the sensors research portfolio, identifying gaps and analyzing advancements in a variety of scientific fields. From 1985 to 1992 he conducted classified research that made pioneering contributions to signature measurement technology and established international standards for radar signature testing. He previously worked in the Passive Observables Branch of the Avionics Laboratory and later in the Air Force Wright Aeronautical Laboratory Signature Technology Office. He is an author of more than 85 archival articles and technical reports and has written key sections of classified textbooks and design manuals. He has delivered more than 200 lectures and developed a DOD Low Observables Short Course that has been taught to more than 2,000 scientists and engineers since 1989. He has served as a technical consultant to a wide range of federal agencies, as an international technical adviser for the DOD, and as a guide for basic research methods to leading academic institutions. He has a BS degree in electrical engineering from Michigan State University and MS and PhD degrees in electrical engineering from Ohio State University.

Dr. Michael A. Kuliasha

Dr. Kuliasha is the chief technologist at the Air Force Research Laboratory (AFRL), where he is the principal technical adviser to the AFRL commander on the Air Force's $2 billion science and technology program and the additional $2 billion customer-funded research and development conducted at AFRL. As chief technologist, he also provides technical guidance for the AFRL workforce of approximately 10,000 people. Prior to his current role, he had more than 30 years of experience in the Department of Energy's Oak Ridge National Laboratory, where he held a variety of leadership positions including director of the Computational Physics and Engineering Division, director for homeland security, and acting associate laboratory director for computing, robotics, and education, among others. He has technical expertise in a broad range of scientific disciplines including nuclear science and technology, high-performance computing, modeling and simulation, knowledge discovery, and energy technologies. He holds a BS degree in mathematics and MS and PhD degrees in nuclear engineering from the University of New Mexico.

Richard Mesic

Mr. Mesic is a senior policy analyst at RAND Corporation with over 39 years of professional experience in requirements definition, system and operational concept development, and quantitative evaluation. His recent work has focused on cyber security; irregular warfare; counterterrorism; and command, control, communications, computers, intelligence, surveillance, and reconnaissance (C4ISR) / space system studies, including information operations and assurance, national intelligence systems, critical infrastructure protection, counter–improvised explosive devices, and special operations. He has studied Air Force operations in all recent conflicts and was principal investigator on the Project Air Force study in support of the Air Force Cyber Command initiative. He continues to colead a multiyear study for the Navy on advanced C4ISR systems for littoral counterterrorism operations. His prior work includes strategic nuclear deterrence and systems, concepts to counter weapons of mass destruction, ballistic missile defense, arms control and arms control verification requirements and capabilities, strategic and tactical C4ISR, national-level intelligence issues, space, information operations, and counterproliferation policy. He has a BA degree in mathematics from Knox College and an MS degree in mathematics from Michigan State University.

Gary A. O'Connell

Mr. O'Connell is chief scientist in the National Air and Space Intelligence Center (NASIC), the Air Force and DOD center of excellence for all-source air and space intelligence. He has over 30 years of experience in aerospace and intelligence fields ranging from electronic countermeasures and tactics development to drafting national intelligence estimates for national policy makers. As chief scientist at NASIC, he is senior advisor to the commander and oversees analytic efforts to support Air Force and joint operational, acquisition, and policy-making customers in the national intelligence community and serves on numerous panels and committees in the national intelligence and scientific communities. He began his career as an air-to-air missile analyst in the Foreign Technology Division and held various management positions until being appointed chief scientist for global threat in 1997. He served as the Air Force representative on the Weapon and Space Systems Intelligence Committee's Aerodynamic Missile Systems Subcommittee and as the Air Force representative on the Scientific and Technical Intelligence Committee from 2001 to 2004. He has a BS degree in aerospace engineering from the University of Cincinnati and an MS degree in national resource strategy from the National Defense University.

Dr. Jim F. Riker

Dr. Riker is the chief scientist of the Space Vehicles Directorate in the Air Force Research Laboratory (AFRL), where he directs technical research in space situational awareness; defensive and offensive space control; responsive space; intelligence, surveillance, and reconnaissance; satellite communications; and satellite control. He has also served in a variety of research roles in optical surveillance and space-based laser system program offices. He is internationally recognized for contributions in the areas of laser beam control, atmospheric propagation, atmospheric compensation, active and passive tracking, and imaging and is an author of more than 50 scientific and professional publications. He previously was technical director for the Optics Division of the Directed Energy Directorate in AFRL and led the Active Track program that produced numerous first-ever technology demonstrations for the Missile Defense Agency's missile defense system. He is a fellow of the Society for Photo-Optical and Instrumentation Engineering and the 2007 winner of the Air Force's Harold M. Brown Award. He has a BS in physics and mathematics from the New Mexico Institute of Technology and an MS and PhD in physics from the University of New Mexico.

Dr. Lara S. Schmidt

Dr. Schmidt is a senior statistician and defense analyst at RAND, where her research focus is military space systems, the Global Positioning

System (GPS), precise time and frequency systems, atomic timekeeping, and risk assessment. Her recent work includes assessments of DOD space systems in hostile environments; space control capabilities; space vulnerability and dependence; weapon system use of space systems; operational level command and control; military positioning, navigation, and timing systems; Air Force special operations; irregular warfare; counterspace threats; and risk to war fighters from attacks on space systems. She is the RAND liaison to Air Force Space Command, is on the advisory board for the DOD precise timing organization, serves as a referee for several technical journals, and has held leadership positions in the American Statistical Association, including chair of the section on defense and national security. Prior to joining RAND, she spent eight years as a government civilian working with GPS and atomic timekeeping. She holds a BS in mathematics from Shepherd College, an MS in mathematics from West Virginia University, and a PhD in mathematics from American University.

Col Eric Silkowski, PhD

Colonel Silkowski is the military assistant to the chief scientist of the US Air Force in the Pentagon. He supports the chief scientist in providing independent, objective, and timely scientific and technical advice to the Air Force chief of staff and the secretary of the Air Force and in evaluating technical issues of relevance to the Air Force mission. He also supports the chief scientist in his contributions to supporting and maintaining the technical quality of the research being conducted across the Air Force. Previously, he led the Air Force Technical Applications Center's Applied Physics Laboratory and ran worldwide operations for nuclear event detection and global atmospheric monitoring. Other technical assignments include high-explosive testing, ballistic reentry vehicle acquisition, and conceptual design of directed-energy weapons systems. He has also served as executive officer to the J8 for NATO Allied Command Operations at Supreme Headquarters Allied Powers, Europe. He has a BA degree in physics from the University of Chicago and has MS and PhD degrees in engineering physics from the Air Force Institute of Technology.

Dr. Dwight C. Streit

Dr. Streit is the vice president for electronics and sensors at Northrop Grumman Aerospace Systems. In this role he is responsible for all technology development required for advanced semiconductors, microelectronics, radars, communication systems and satellite payload electronics, and sensors and electronic systems for airborne and space-based platforms. Prior to joining Northrop Grumman via its acquisition of TRW, he was president of the TRW company that manufactured high-performance chips for fiber-optic and wireless communication systems. He has contributed more than 300 technical publications and conference presentations and has some 30 patents issued or pending. He was made a member of the National Academy of Engineering for his contributions to the development and production of heterojunction transistors and circuits. He is a fellow of the Institute of Electrical and Electronics Engineers, a member of the NASA Space Foundation Technology Hall of Fame, and a fellow of the American Association for the Advancement of Science. In 2003 he was named the University of California, Los Angeles (UCLA) Engineering Alumnus of the Year. He has BS degrees in engineering and chemistry from California State University and MS and PhD degrees in electrical engineering from UCLA.

Betsy S. Witt

Ms. Witt is technical director of the Air and Cyberspace Analysis Group of the National Air and Space Intelligence Center, the Air Force and DOD center of excellence for all-source air and space intelligence to support USAF and joint operations, acquisition, and policy-making customers in the national intelligence community. The Air and Cyberspace Analysis Group is the intelligence community focal point for analyses of foreign air and air defense systems and for analyses of cyber intrusions into Air Force computer networks. In her capacity as technical director, she provides technical oversight of analysis on foreign fixed-wing fighters and air armaments, remotely piloted vehicles, aircraft avionics, ground-based early warning radars, command and control, integrated air defense systems, and foreign computer network attack. She began her career in the Foreign Technology Division and has over 31 years' experience in intelligence community assignments with the Air Force. She has BS and MS degrees in mathematics from Wright State University, as well as extensive undergraduate and graduate-level courses in engineering.

Dr. Marc A. Zissman

Dr. Zissman is the assistant head of the Communications and Information Technology Division at the Massachusetts Institute of Technology (MIT) Lincoln Laboratory. He leads the laboratory's cyber protection activities and shares responsibility for research, development, evaluation, and technology transfer in the areas of advanced military communications, wideband tactical networking on the move, and language processing. He joined the laboratory in 1983, with his early research focused on digital speech processing, including parallel computing for speech coding and recognition, co-channel talker interference suppression, language and dialect identification, and cochlear-implant processing. After his subsequent research interests expanded to include information assurance technologies, he served for four years as a US technical specialist to the NATO task group that studies military applications of speech technology and for four years on the Speech Processing Technical Committee of the Institute of Electrical and Electronics Engineers Signal Processing Society. He also served four years on the Defense Advanced Research Projects Agency Information Science and Technology Study Group. He holds SB, SM, and PhD degrees in electrical engineering from MIT.

Dr. John C. Zolper

Dr. Zolper is vice president, corporate research and development, and deputy for corporate technology and research for Raytheon Company. In this role he is responsible for strategic technology planning and technology innovation across the enterprise, through which he has been involved in development and execution of an integrated enterprise-wide technology and research vision and strategy. Before joining Raytheon he was the director of the Defense Advanced Research Projects Agency's (DARPA) Microsystems Technology Office, where he was responsible for the strategic planning and execution of research programs covering all areas of advanced component technology, including electronics, photonics, microelectromechanical systems, algorithms, and component architecture. Prior to joining DARPA he was a program officer at the Office of Naval Research, where he managed a portfolio of basic and applied research in advanced electronics, was a senior member of technical staff at Sandia National Laboratories, and was a postdoctoral fellow conducting research on high-efficiency silicon solar cells at the University of New South Wales in Australia. He is a fellow of the Institute of Electrical and Electronics Engineers and the American Institute of Aeronautics and Astronautics. He has a BA in physics from Gettysburg College and MEE and PhD degrees in electrical engineering from the University of Delaware.

Abbreviations

AAM	air-to-air missile
ACC	Air Combat Command
AESA	active electronically scanned array
AFGSC	Air Force Global Strike Command
AFOSR	Air Force Office of Scientific Research
AFRL	Air Force Research Laboratory
AFSOC	Air Force Special Operations Command
AFSPC	Air Force Space Command
AMC	Air Mobility Command
AOARD	Asian Office of Aerospace Research and Development
ASAT	antisatellite
CBO	Congressional Budget Office
CEP	circular error probable
DARPA	Defense Advanced Research Projects Agency
DE	directed energy
DOD	Department of Defense
DRFM	digital radio frequency memory
EELV	evolved expendable launch vehicle
EM	electromagnetic
EMP	electromagnetic pulse
EOARD	European Office of Aerospace Research and Development
EO-IR	electro-optical-infrared
EW	electronic warfare
FFRDC	federally funded research and development center
GDP	gross domestic product
GEO	geosynchronous Earth orbit
GLONASS	Global Navigation Satellite System
GPS	Global Positioning System
GRAM	GPS receiver application module
HAF	Headquarters Air Force
HALE	high-altitude long-endurance
HELLADS	High-Energy Liquid Laser Area Defense System

HEO	highly elliptical orbit
IADS	integrated air defense system
IARPA	Intelligence Advanced Research Projects Agency
ICBM	intercontinental ballistic missile
IEEE	Institute of Electrical and Electronics Engineers
IMU	inertial measurement unit
INS	inertial navigation system
IP	Internet protocol
IRST	infrared search and track
ISR	intelligence, surveillance, and reconnaissance
JPADS	joint precision airdrop system
kg	kilogram
KTA	key technology area
LEO	low Earth orbit
LIDAR	light detection and ranging
LO	low observable
M&S	modeling and simulation
MANPADS	man-portable air defense system
MDA	MacDonald, Dettwiler and Associates
MEO	medium Earth orbit
MMOG	massively multiplayer online game
MRBM	medium-range ballistic missile
NASA	National Aeronautics and Space Administration
OPR	overall pressure ratio
PCA	potential capability area
PED	processing, exploitation, and dissemination
PLAAF	People's Liberation Army Air Force
PNT	positioning, navigation, and timing
PRSEUS	pultruded, rod-stitched efficient unitized structure
QKD	quantum key distribution
R&D	research and development
RF	radio frequency

RPA	remotely piloted aircraft
S&T	science and technology
SAM	surface-to-air missile
SBSS	space-based space surveillance
SCF	service core function
SOARD	Southern Office of Aerospace Research and Development
SSA	space situational awareness
SST	space surveillance telescope
STEM	science, technology, engineering, and mathematics
T&E	test and evaluation
TPS	thermal protection system
TRL	technology readiness level
TSTO	two stage to orbit
V&V	verification and validation

Bibliography

Studies, Reports, and Other Documents

Aeronautics Research, Development, Test, and Evaluation (RDT&E) Infrastructure Interagency Working Group. "National Aeronautics RDT&E Infrastructure Plan: Final IIWG Draft." Aeronautics RDT&E Infrastructure Interagency Working Group, 2009.

Air University. *Air Force 2025*. A future study conducted 1995–96 for the Air Force chief of staff. Maxwell AFB, AL: Air Education and Training Command, 1996.

———. *Spacecast 2020: Air University into the Future*. A future study conducted 1993–94 for the Air Force chief of staff. Maxwell AFB, AL: Air Education and Training Command, 1994.

Air University Center for Strategy and Technology. *Blue Horizons 2007: "Horizon 21" Project Report*. Maxwell AFB, AL: Air University Center for Strategy and Technology, 2007.

Alexander, Keith B. "Warfighting in Cyberspace." *Joint Force Quarterly* 46 (2007): 58–61.

Amouzegar, Mahyar A., Ronald G. McGarvey, Robert S. Tripp, Louis Luangkesorn, Thomas Lang, and Charles Robert Roll, Jr. *Evaluation of Options for Overseas Combat Support Basing*. RAND Project Air Force. Santa Monica, CA: RAND, 2006.

Andrews, Amy E., and Mark E. Segal. "An Overview of Cloud Computing." *The Next Wave: The National Security Agency's Review of Emerging Technology* 17, no. 4 (2009): 6–18.

Arthur, David. *Options for Strategic Military Transportation Systems*. Washington, DC: Congressional Budget Office, 2005.

Assistant Secretary of the Air Force for Installations, Environment, and Logistics. *Air Force Energy Plan 2010*. Washington, DC: Office of the Secretary of the Air Force, Headquarters Air Force, 2010.

Association for Computing Machinery. *Proceedings of the 1st Augmented Human International Conference*. Megève, France, 2–3 April 2010. New York: Association for Computing Machinery, 2010.

Australian Department of Defence. *Defending Australia in the Asia Pacific Century: Force 2030*. Defence White Paper. Canberra, Australia: Department of Defence, 2009.

Beason, Doug. *DOD Science and Technology: Strategy for the Post–Cold War Era*. Washington, DC: National Defense University Press, 1997.

———. *The E-Bomb: How America's New Directed Energy Weapons Will Change the Way Future Wars Will Be Fought*. Cambridge, MA: Da Capo Press, 2005.

Bell, Gordon, Jim Gray, and Alex Szalay, Microsoft Research and Johns Hopkins University. "Petascale Computational Systems: Balanced Cyber Infrastructure in a Data-Centric World." *Computer* 39, no. 1 (January 2006): 110–12.

Bilar, Daniel, Department of Computer Science, University of New Orleans. "On n^{th} Order Attacks." *Proceedings of the Conference on Cyber Warfare 2009*. Fairfax, VA: IOS Press, 2009.

Blank, Stephen. *Rethinking Asymmetric Threats*. Carlisle Barracks, PA: Strategic Studies Institute, US Army War College, 2003.

Bowie, Christopher. *The Anti-Access Threat and Theater Air Bases*. Washington, DC: Center for Strategic and Budgetary Assessments, 2002.

Brockman, John, ed. *The Next Fifty Years: Science in the First Half of the Twenty-First Century*. New York: Vintage Press, 2002.

Brown, Col Lex, and Lt Col Anthony P. Tvaryanas. "Human Performance Enhancement: Überhumans or Ethical Morass?" *Air and Space Power Journal* 22, no. 4 (Winter 2008): 39–43.

Bush, George W. *The National Security Strategy of the United States of America*. Washington, DC: The White House, 2006.

Calandrino, Joseph A., Ariel J. Feldman, Jacob Appelbaum, and Edward W. Felten. "Lest We Remember: Cold Boot Attacks on Encryption Keys." *Proceedings 2008 USENIX Security Symposium*. Berkeley, CA: USENIX Association, 2008.

Cameron, Rebecca, and Barbara Wittig, eds. *Golden Legacy, Boundless Future: Essays on the United States Air Force and the Rise of Aerospace Power: Proceedings of a Symposium Held on May 28–29, 1997, at the Double Tree Hotel, Crystal City, Virginia*. Washington, DC: Air Force History and Museums Program, 2000.

Carin, Lawrence, George Cybenko, and Jeff Hughes. "Cybersecurity Strategies: The QuERIES Methodology." *Computer* 41, no. 8 (August 2008): 20.

Catlett, Charlie, ed. *A Scientific Research and Development Approach to Cyber Security*. Report submitted to the Department of Energy on behalf of the Research and Development Community. Argonne, IL: Argonne National Laboratory, 2008.

Center for Strategy and Technology (CSAT). "Operational Impact of Exponential Technological Change on the Air Force." Slide presentation in support of CSAT's *Blue Horizons 2007* report. Maxwell AFB, AL: Center for Strategy and Technology, Air University, 2008.

Clodfelter, Mark. "Back from the Future: The Impact of Change on Airpower in the Decades Ahead." *Strategic Studies Quarterly* 3, no. 3 (Fall 2009): 104–22.

Cloud Security Alliance. "Security Guidance for Critical Areas of Focus in Cloud Computing." Version 2.1. Cloud Security Alliance, December 2009. Accessed 18 July 2011. https://cloudsecurityalliance .org/csaguide.pdf.

Commandant of the Marine Corps, Chief of Naval Operations, and Commandant of the Coast Guard. *A Cooperative Strategy for 21st Century Seapower*. Washington, DC: US Marine Corps, US Dept. of the Navy, US Coast Guard, 2007.

Congressional Budget Office. *Alternatives for Future U.S. Space-Launch Capabilities*. Washington, DC: Congressional Budget Office, 2006.

———. *Alternatives for Long-Range Ground-Attack Systems*. Washington, DC: Congressional Budget Office, 2006.

———. *The Long-Term Implications of Current Plans for Investment in Major Unclassified Military Space Programs*. Washington, DC: Congressional Budget Office, 2005.

Congressional Commission on the Strategic Posture of the United States. *America's Strategic Posture: The Final Report of the Congressional Commission on the Strategic Posture of the United States*. Washington, DC: United States Institute of Peace Press, 2009.

Council on Competitiveness. *Dialogue 1: The Changing Global Landscape for Technology Leadership*. Washington, DC: Council on Competitiveness, 2009.

Cragin, Kim, and Scott Gerwehr. *Dissuading Terror: Strategic Influence and the Struggle against Terrorism*. Santa Monica, CA: RAND, 2005.

Daso, Dik A. *Architects of American Air Supremacy: Gen Hap Arnold and Dr. Theodore von Kármán*. Maxwell AFB, AL: Air University Press, 1997.

Day, Dwayne A. *Lightning Rod: A History of the Air Force Chief Scientist's Office*. Washington, DC: Chief Scientist's Office, US Air Force, 2000.

Defense Advanced Research Projects Agency. "Cold Atom Inertial Navigation Technology." Defense Science Office, 2010.

Defense Science Board. *Challenges to Military Operations in Support of U.S. Interests*. Vol. 1, *Executive Summary*. Vol. 2, *Main Report*. Washington, DC: Office of the Under Secretary of Defense for Acquisition, Technology, and Logistics, 2008.

———. *Creating an Assured Joint DOD and Interagency Interoperable Net-Centric Enterprise*. Washington, DC: Defense Science Board, 2009.

———. *Future Need for VTOL [Vertical Take-Off and Landing]/STOL [Short Take-Off and Landing] Aircraft*. Washington, DC: Office of the Under Secretary of Defense for Acquisition, Technology, and Logistics: Defense Science Board, 2007.

———. *The Militarily Critical Technologies List*. Report no. ASDR-169. Washington, DC: Office of the Under Secretary of Defense for Acquisition, Technology, and Logistics, 2010.

———. *Operations Research Applications for Intelligence Surveillance and Reconnaissance (ISR)*. Washington, DC: Defense Science Board, 2009.

———. *Report of the Defense Science Board Task Force on Strategic Communication*. Washington, DC: Office of the Under Secretary of Defense for Acquisition, Technology, and Logistics, 2008.

———. *Report of the Defense Science Board Task Force on Time Critical Conventional Strike for Strategic Standoff*. Washington, DC: Office of the Under Secretary of Defense for Acquisition, Technology, and Logistics, 2009.

———. *Report of the Defense Science Board Task Force on Understanding Human Dynamics*. Washington, DC: Defense Science Board, 2009.

———. *Report of the Defense Science Board 2008 Summer Study on Capability Surprise*. 2 vols. Washington, DC: Office of the Under Secretary of Defense for Acquisition, Technology, and Logistics, 2009–10.

———. *21st Century Strategic Technology Vectors.* Vol. 1, *Main Report.* Vol. 2, *Critical Capabilities and Enabling Technologies.* Vol. 3, *Strategic Technology Planning.* Vol. 4, *Accelerating the Transition of Technologies into U.S. Capabilities.* Washington, DC: Office of the Under Secretary of Defense for Acquisition, Technology, and Logistics, 2007.

———. *Wideband Radio Frequency Modulation: Dynamic Access to Mobile Information Networks.* Washington, DC: Defense Science Board, 2003.

Department of the Air Force. *Air Force Roadmap 2006–2025.* Washington, DC: US Air Force, 2006.

———. *Global Reach—Global Power: The Evolving Air Force Contribution to National Security.* White paper. Report no. DTIC 19970610 059. Washington, DC: Aerospace Education Foundation, 1992.

———. *Lead Turning the Future: The 2008 Strategy for United States Air Force Intelligence, Surveillance and Reconnaissance.* Washington, DC: Headquarters, United States Air Force, 2008.

Department of the Army. *Army Energy Security Implementation Strategy.* Washington, DC: Army Senior Energy Council and Office of the Deputy Assistant Secretary of the Army for Energy and Partnerships, 2009.

Dinda, Peter. "Addressing the Trust Asymmetry Problem in Grid Computing with Encrypted Computation." *Proceedings of the Seventh Workshop of Languages, Compilers and Run-Time Support for Scalable Systems (LCR 2004).* New York: Association for Computing Machinery, Inc., 2004.

Drew, Dennis M., and Donald M. Snow. *Making Twenty-First-Century Strategy: An Introduction to Modern National Security Processes and Problems.* Maxwell AFB, AL: Air University Press, 2006.

Echevarria, Antulio. *Fourth-Generation War and Other Myths.* Carlisle Barracks, PA: Strategic Studies Institute, US Army War College, 2005.

Ehrhard, Thomas. *An Air Force Strategy for the Long Haul.* Strategy for the Long Haul Series. Washington, DC: Center for Strategic and Budgetary Assessments, 2009.

Ewart, Roberta, chief scientist, Space and Missile Systems Center. "Technologies of Choice for Future Space Investment." Briefing slides, 13 March 2009.

Executive Office of the President, National Science and Technology Council. *National Plan for Aeronautics Research and Development and Related Infrastructure.* Washington, DC: National Science and Technology Council, 2007.

Friedman, George. *The Future of War: Power, Technology, and American World Dominance in the 21st Century.* New York: St. Martin's Griffin, 1998.

———. *The Next 100 Years: A Forecast for the 21st Century.* New York: Doubleday, 2009.

Friedman, Thomas. *The World Is Flat: A Brief History of the Twenty-First Century.* New York: Picador, 2007.

Gates, Robert M. *Quadrennial Defense Review Report.* Washington, DC: Department of Defense, 2010.

Gorn, Michael H. *Harnessing the Genie: Science and Technology Forecasting for the Air Force, 1944–1986.* Washington, DC: Office of Air Force History, 1988.

———. *Prophecy Fulfilled: "Toward New Horizons" and Its Legacy.* Bolling AFB, DC: Air Force History and Museums Program, 1994.

Gouré, Daniel, and Christopher M. Szara, eds. *Air and Space Power in the New Millennium.* Washington, DC: Center for Strategic and International Studies, 1997.

Grant, Rebecca. *Return of the Bomber: The Future of Long-Range Strike.* Arlington, VA: Air Force Association, 2007.

Gray, Colin S. *Modern Strategy.* New York: Oxford University Press, 1999.

———. *Transformation and Strategic Surprise.* Carlisle Barracks, PA: Strategic Studies Institute, US Army War College, 2005.

Grimmett, Richard. *Conventional Arms Transfers to Developing Nations: 2001-2008.* Washington, DC: Congressional Research Service, 2009.

Hazell, J. Eric. *From Reform to Reduction: Reports on the Management of Navy and Department of Defense Laboratories in the Post–Cold War Era.* Washington, DC: National Defense University and Naval Historical Center, 2008.

Headquarters Air Force. *U.S. Air Force Intelligence, Surveillance, and Reconnaissance (ISR) Flight Plan: The Key Guidance for Planning the Current and Future ISR Force.* Washington, DC: Deputy Chief of Staff ISR (AF/A2), 2009.

Higham, Robin D. S., and Stephen John Harris, eds. *Why Air Forces Fail: The Anatomy of Defeat*. Lexington, KY: University of Kentucky Press, 2006.

Holland, O. Thomas. "Taxonomy for the Modeling and Simulation of Emergent Behavior Systems." *SpringSim '07: Proceedings of the 2007 Spring Simulation Multiconference*. Vol. 2. San Diego, CA: Society for Computer Simulation International, 2007.

Hull, George. "Security and Complexity: Are We on the Wrong Road?" In *CSIIRW 2009 Proceedings of the 5th Annual Workshop on Cyber Security and Information Intelligence Research: Cyber Security and Information Intelligence Challenges and Strategies*. New York: ACM, 2009.

IBM. *Seeding the Clouds: Key Infrastructure Elements for Cloud Computing*. Somers, NY: IBM, 2009.

Jacobs, Jody. *Technologies and Tactics for Improved Air-Ground Effectiveness*. Report no. MG-573-AF. RAND Project Air Force. Santa Monica, CA: RAND, 2008.

Johns, Lt Gen Ray, Headquarters Air Force. "Air Force Strategy to Resources." Briefing slides, 9 December 2008.

———. *US Air Force Futures Capabilities Game 2007: Unclassified Report Summary*. Washington, DC: Directorate of Strategic Planning, 2008.

Johnson, Stephen. *The United States Air Force and the Culture of Innovation: 1945–1965*. Washington, DC: Air Force History and Museums Program, 2002.

Karagozian, Ann, Werner Dahm, Ed Glasgow, Roger Howe, Ilan Kroo, Richard Murray, and Heidi Shyu. *Technology Options for Improved Air Vehicle Fuel Efficiency: Executive Summary and Annotated Brief*. Report no. SAB-TR-06-04. Washington, DC: USAF Scientific Advisory Board, 2006.

Kelly, Henry, Ivan Oelrich, Steven Aftergood, and Benn H. Tannenbaum. *Flying Blind: The Rise, Fall, and Possible Resurrection of Science Policy Advice in the United States*. Occasional paper no. 2. Washington, DC: Federation of American Scientists, 2004.

Kent, Glenn, David Ochmanek, Michael Spirtas, and Bruce Pirnie. *Thinking about America's Defense: An Analytical Memoir*. RAND Project Air Force. Santa Monica, CA: RAND, 2008.

Kilcullen, David. *The Accidental Guerrilla: Fighting Small Wars in the Midst of a Big One*. New York: Oxford University Press, 2009.

Kitching, John. *Chip-Scale Atomic Devices: Precision Instruments Based on Lasers, Atoms and MEMS*. Boulder, CO: National Institute of Standards and Technology, 2009.

Kopp, Carlo. "China's Air Defence Missile Systems." *Defence Today*, March/April 2008, 22–24.

———. "Post Cold War Air-to-Air Missile Evolution." *Defence Today*, March 2009, 56–59.

Kosiak, Steven M. *Matching Resources with Requirements: Options for Modernizing the US Air Force*. Washington, DC: Center for Strategic and Budgetary Assessments, 2004.

———. *Military Compensation: Requirements, Trends and Options*. Washington, DC: Center for Strategic and Budgetary Assessments, 2005.

———. *Spending on US Strategic Nuclear Forces: Plans and Options for the 21st Century*. Washington, DC: Center for Strategic and Budgetary Assessments, 2006.

Krekel, Bryan. *Capability of the People's Republic of China to Conduct Cyber Warfare and Computer Network Exploitation*. McClean, VA: Information Systems, Northrop Grumman, 2009.

Krepinevich, Andrew. *Seven Deadly Scenarios: A Military Futurist Explores War in the 21st Century*. New York: Bantum Books, 2009.

Krepinevich, Andrew, Barry Watts, and Robert Work. *Meeting the Anti-Access and Area-Denial Challenge*. Washington, DC: Center for Strategic and Budgetary Assessments, 2003.

Krepinevich, Andrew, Robert Martinage, and Robert Work. *The Challenges to US National Security*. Strategy for the Long Haul Series. Washington, DC: Center for Strategic and Budgetary Assessments, 2008.

Larson, Eric. *Assuring Access in Key Strategic Regions: Toward a Long-Term Strategy*. Santa Monica, CA: RAND, 2004.

Li, V., and A. Velicki. "Advanced PRSEUS Structural Concept Design and Optimization." Presentation. American Institute of Aeronautics and Astronautics, British Columbia, Canada, September 2008.

Libicki, Martin C. *Cyberdeterrence and Cyberwar*. RAND Project Air Force. Santa Monica, CA: RAND, 2009.

Lin, Patrick, George Bekey, and Keith Abney. *Autonomous Military Robotics: Risk, Ethics, and Design*. Arlington, VA: Office of Naval Research, 2008.

Long, Austin. *On "Other War": Lessons from Five Decades of RAND Counterinsurgency Research*. Santa Monica, CA: RAND Corporation, 2006.

MacDonald, Bruce. *China, Space Weapons, and US Security*. New York: Council on Foreign Relations, September 2008.

Mackey, Lt Col J. "Recent US and Chinese Antisatellite Activities." *Air and Space Power Journal* 23, no. 9 (2009): 82–93.

Mahnken, Thomas. *The Cruise Missile Challenge*. Washington, DC: Center for Strategic and Budgetary Assessments, 2005.

Martino, Maj J. "Forecasting the Progress of Technology." *Air University Review* 20, no. 3 (March–April 1969): 11–20.

Marvin, Dean C., Aerospace Corporation. "Game Changer Concepts." Briefing slides, May 2007.

Mendel, William, and Murl Munger. *Strategic Planning and the Drug Threat*. Carlisle Barracks, PA: Strategic Studies Institute, US Army War College, 1997.

Mesic, Richard, David E. Thaler, David Ochmanek, and Leon Goodson. *Courses of Action for Enhancing U.S. Air Force "Irregular Warfare" Capabilities: A Functional Solutions Analysis*. RAND Project Air Force. Santa Monica, CA: RAND Corporation, 2010.

Metz, Steven, and Raymond Millen. *Insurgency and Counterinsurgency in the 21st Century: Reconceptualizing Threat and Response*. Carlisle Barracks, PA: Strategic Studies Institute, US Army War College, 2004.

Ministry of Defence, United Kingdom. *Defence Technology Strategy for the Demands of the 21st Century*. UK: Ministry of Defence, 2006.

———. *Strategic Trends Programme: Future Character of Conflict*. UK: Ministry of Defence, 2010.

Moseley, Gen T. Michael. *The Nation's Guardians: America's 21st Century Air Force*. Chief of Staff of the Air Force White Paper. Washington, DC: Dept. of the Air Force, Office of the Chief of Staff, 2008.

Murray, Williamson, ed. *Strategic Challenges for Counterinsurgency and the Global War on Terrorism*. Carlisle Barracks, PA: Strategic Studies Institute, US Army War College, 2006.

National Intelligence Council (NIC). *Global Trends 2025: A Transformed World*. Washington, DC: NIC, 2008.

———. *Mapping the Global Future: Report of the National Intelligence Council's 2020 Project*. Washington, DC: NIC, 2004.

National Research Council. *Avoiding Surprise in an Era of Global Technology Advances / Committee on Defense Intelligence Agency Technology Forecasts and Review, Division on Engineering and Physical Sciences, National Research Council*. Washington, DC: National Academies Press, 2005.

———. *Effectiveness of Air Force Science and Technology Program Changes / Committee on Review of the Effectiveness of Air Force Science and Technology Program Changes; Air Force Science and Technology Board; Division on Engineering and Physical Sciences; National Research Council of the National Academies*. Washington, DC: National Academies Press, 2003.

———. *Future Air Force Needs for Survivability / Committee on Future Air Force Needs for Survivability, Air Force Studies Board, Division on Engineering and Physical Sciences, National Research Council of the National Academies*. Washington, DC: National Academies Press, 2006.

———. *A Review of United States Air Force and Department of Defense Aerospace Propulsion Needs / Committee on Air Force and Department of Defense Aerospace Propulsion Needs, Air Force Studies Board, Division on Engineering and Physical Sciences, National Research Council of the National Academies*. Washington, DC: National Academies Press, 2006.

National Science Board. *Science and Engineering Indicators 2008*. 2 vols. Arlington, VA: National Science Foundation, 2008.

———. *Science and Engineering Indicators 2010*. Arlington, VA: National Science Foundation, 2010.

Neufeld, Jacob, ed. *Guideposts for the United States Military in the Twenty-First Century: Symposium Proceedings, September 16-17, 1999, Bolling Air Force Base, Washington, D.C. / Co-Sponsored by the Air Force Historical Foundation and the Office of the Air Force Historian, with the Support of the McCormick Tribune Foundation*. Washington, DC: Air Force History and Museums Program, 2000.

Nguyen, Clark, and John Kitching. "Towards Chip-Scale Atomic Clocks." Research paper. Arlington, VA: Defense Advanced Research Projects Agency, Microsystems Technology Office, 2005.

Ochmanek, David. *Military Operations against Terrorist Groups Abroad: Implications for the United States Air Force.* RAND Project Air Force. Santa Monica, CA: RAND, 2003.

Office of the Chief of Naval Operations (CNO), Strategic Studies Group XXVIII. *The Unmanned Imperative.* Newport, RI: CNO Strategic Studies Group, 2009.

Office of the Director of Defense Research and Engineering, Joint Chiefs of Staff, and Department of Defense. *Joint Warfighting Science and Technology Plan.* Washington, DC: DOD, 2008.

Orrell, David. *The Future of Everything: The Science of Prediction.* New York: Thunder's Mouth Press, 2007.

Post, Joseph, and Michael Bennett. *Alternatives for Military Space Radar.* Washington, DC: Congressional Budget Office, 2007.

Pu, Calton. "Infosphere: Smart Delivery of Fresh Information." Presentation slides. Georgia Tech, 2000. http://www.cc.gatech.edu/projects /infosphere/SBBDkeynote/ppframe.htm

Pumphrey, Carolyn, ed. *Transnational Threats: Blending Law Enforcement and Military Strategies.* Carlisle Barracks, PA: Strategic Studies Institute, US Army War College, 2000.

Rees, Dr. William S., Jr., Deputy Undersecretary of Defense (Laboratories and Basic Sciences). "Why Should DoD Invest in Basic Research?" Presentation. 33d Annual Government Microcircuit Applications and Critical Technology Conference (GOMAC), Las Vegas, NV, March 2008.

Roco, Mihail C., and Williams Sims Bainbridge, eds. *Converging Technologies for Improving Human Performance: Nanotechnology, Biotechnology, Information Technology and Cognitive Science.* Arlington, VA: National Science Foundation & Department of Commerce, 2002.

Scales, Maj Gen Robert, Jr., ed. *Future Warfare Anthology.* Rev. ed. Carlisle Barracks, PA: Strategic Studies Institute, US Army War College, 2000.

Schneider, William, Jr. *Defense Science Board Task Force on Directed Energy Weapons.* Washington, DC: Defense Science Board, 2007.

Schwartz, Gen Norton A., and Lt Col Timothy R. Kirk. "Policy and Purpose: The Economy of Deterrence." *Strategic Studies Quarterly* 3, no. 1 (Spring 2009): 11–30.

Shlapak, David. *Shaping the Future Air Force*. RAND Project Air Force. Santa Monica, CA: RAND, 2006.

Silberglitt, Richard, Philip Anton, David Howell, Anny Wong, Natalie Gassman, Brian Jackson, Eric Landree, Shari Lawrence Pfleeger, Elaine Newton, and Felicia Wu. *The Global Technology Revolution 2020, In-Depth Analyses: Bio/Nano/Materials/Information Trends, Drivers, Barriers, and Social Implications*. Santa Monica, CA: RAND, 2006.

Singer, Peter. *Wired for War: The Robotics Revolution and Conflict in the 21st Century*. New York: Penguin Press, 2009.

Smith, Maj Ron, and Scott Knight. "Applying Electronic Warfare Solutions to Network Security." *Canadian Military Journal* 6 (Autumn 2005): 49–58.

Stillion, John, and Scott Perdue. *Air Combat Past, Present and Future*. RAND Project Air Force. Santa Monica, CA: RAND, 2008.

Stytz, Martin, and Jeff Hughes, AFRL, Wright-Patterson AFB, OH. "Software Protection Initiative: Software Application Security." Presentation slides, 30 June 2003.

Thomas, Timothy L. "Chinese and American Network Warfare." *Joint Force Quarterly* 38 (2005): 76–83.

USAF Scientific Advisory Board. *Air Defense against Unmanned Aerial Vehicles*. Vol. 1, *Executive Summary and Annotated Brief*. Vol. 2, *Final Report*. Report no. SAB-TR-06-01. Washington, DC: USAF Scientific Advisory Board, 2006.

———. *Implications of Cyber Warfare*. Vol. 1, *Executive Summary and Annotated Brief*. Vol. 2, *Main Report*. Washington, DC: USAF Scientific Advisory Board, 2007.

———. *Implications of Spectrum Management for the Air Force*. Report no. SAB-TR-08-03. Washington, DC: USAF Scientific Advisory Board, 2008.

———. *New World Vistas: Air and Space Power for the 21st Century*. 13 vols. Washington, DC: USAF Scientific Advisory Board, 1995.

———. *System-Level Experimentation: Executive Summary and Annotated Brief.* Report no. SAB-TR-06-02. Washington, DC: USAF Scientific Advisory Board, 2006.

———. *Thermal Management Technology Solutions.* Report no. SAB-TR-07-05. Washington, DC: USAF Scientific Advisory Board, 2007.

———. *Use and Sustainment of Composites in Aircraft.* USAF Scientific Advisory Board, SAB-TR-07-04, 2007.

US Air Force Futures Group. To the chief of staff of the Air Force. Memorandum. "Strategic Concerns That Deserve CSAF Attention," 2007.

US Joint Forces Command. *The JOE, Joint Operating Environment, 2010.* Norfolk, VA: US Joint Forces Command, 2010.

Velicki, Alex, and Patrick Thrash. "Advanced Structural Concept Development Using Stitched Composites." Presentation AIAA 2008-2329. American Institute of Aeronautics and Astronautics Conference, Schaumburg, IL, 7–10 April 2008.

Vick, Alan J., Adam Grissom, William Rosenau, Beth Grill, and Karl P. Mueller. *Air Power in the New Counterinsurgency Era: The Strategic Importance of USAF Advisory and Assistance Missions.* RAND Project Air Force. Santa Monica, CA: RAND, 2006.

Vick, Alan J., Richard M. Moore, Bruce R. Pirnie, and John Stillion. *Aerospace Operations against Elusive Ground Targets.* Report no. MR-1398-AF. RAND Project Air Force. Santa Monica, CA: RAND Corporation, 2001.

Vickers, Michael, and Robert Martinage. *Future Warfare 20XX Wargame Series: Lessons Learned Report.* Washington, DC: Center for Strategic and Budgetary Assessments, 2001.

———. *The Revolution in War.* Washington, DC: Center for Strategic and Budgetary Assessments, 2004.

Von Kármán, Theodore, ed. *Toward New Horizons: A Report to the General of the Army H. H. Arnold, Submitted on Behalf of the Army Air Forces Scientific Advisory Group.* Wright Field, OH: Air Materiel Command Publications Branch, Intelligence, T-2, 1945.

Waters, Gary, Desmond Ball, and Ian Dudgeon. *Australia and Cyber-Warfare.* Canberra, Australia: ANU E Press, 2008.

Watts, Barry. *The Military Use of Space: A Diagnostic Assessment.* Washington, DC: Center for Strategic and Budgetary Assessments, 2001.

Williams, E., et al. *Human Performance.* JASON Report JSR-07-625. McLean, VA: MITRE Corporation, 2008.

Zhang, Xiaoming, and Col Sean D. McClung. "The Art of Military Discovery: Chinese Air and Space Power Implications for the USAF." *Strategic Studies Quarterly* 4, no. 1 (Spring 2010): 36–62.

Site Visits, Briefings, and Discussions

Technology Horizons also made use of a broad range of inputs gained from numerous site visits, briefings, and discussions involving organizations within the Air Force and elsewhere in the Department of Defense, as well as other federal agencies, federally funded research and development centers, national laboratories, and companies. A partial list of these organizations is given below.

Air Staff, Headquarters Air Force (HAF)

- Manpower and Personnel (AF/A1)
- Intelligence, Surveillance, and Reconnaissance (AF/A2)
- Operations, Plans, and Requirements (AF/A3/5)
- Logistics, Installations, and Mission Support (AF/A4/7)
- Strategic Plans and Programs (AF/A8)
- Studies, Assessments, and Lessons Learned (AF/A9)
- Strategic Deterrence and Nuclear Integration (AF/A10)
- Air Force Historian (AF/HO)
- Air Force Reserve (AF/RE)
- Air Force Safety (AF/SE)
- Air Force Chief Scientist (AF/ST)
- Air Force Surgeon General (AF/SG)
- Air Force Test and Evaluation (AF/TE)
- Air Force Scientific Advisory Board (AF/SB)
- CSAF Strategic Studies Group

Air Force Secretariat, HAF

- Administrative Assistant (SAF/AA)
- Acquisition (SAF/AQ)
- Small Business Programs (SAF/SB)
- General Counsel (SAF/GC)

- Installations, Environment, and Logistics (SAF/IE)
- Public Affairs (SAF/PA)
- Warfighting Integration and Chief Information Officer (SAF/XC)

Air Force Major Commands (MAJCOM)

- Air Combat Command (ACC)
- Air Force Materiel Command (AFMC)
- Air Force Space Command (AFSPC)
- Air Education and Training Command (AETC)
- Air Force Reserve Command (AFRC)
- Air Force Special Operations Command (AFSOC)
- Air Mobility Command (AMC)
- Air National Guard (ANG)
- Pacific Air Forces (PACAF)

Air Force Product Centers

- Air Armament Center (AAC)
- Electronic Systems Center (ESC)
- Space and Missile Systems Center (SMC)

Direct Reporting Units (DRU)

- US Air Force Academy (USAFA)

Field Operating Agencies (FOA)

- Air Force ISR Agency (AFISRA)
- Air Force Medical Support Agency (AFMSA)

Other Air Force Agencies

- Air Force Arnold Engineering Development Center (AEDC)
- Air Force Flight Test Center (AFFTC)
- Air Force Institute of Technology (AFIT)
- Air Force Research Laboratory (AFRL)
- Air Force Unmanned Systems ISR (Intelligence, Surveillance, and Reconnaissance) Innovations (AF/A2U)
- Air Force Warfare Center (USAFWC)
- Air Force Nuclear Weapons Center (AFNWC)
- Air Force A8 Futures Game 2009

- Distributed Mission Operations Center (DMOC)
- Space Innovation and Development Center (SIDC)
- National Air and Space Intelligence Center (NASIC)
- Oklahoma City Air Logistics Center (OC–ALC)
- Eighth Air Force (8AF)
- 505th Command and Control Wing (505CCW)
- 688th Information Operations Wing (688IOW)
- 67th Network Warfare Wing (67NWW)
- 46th Test Wing (46TW)

Federal Agencies

- Defense Advanced Research Projects Agency (DARPA)
- Defense Threat Reduction Agency (DTRA)
- National Aeronautics and Space Administration (NASA)
- Office of Science and Technology Policy (OSTP)

Federally Funded Research and Development Centers (FFRDC)

- RAND Corporation
- Lincoln Laboratory

National Laboratories

- Argonne National Laboratory
- Lawrence Livermore National Laboratory
- Sandia National Laboratories

Companies

- Aernnova
- Aerojet
- Astrox
- Blue Origin
- Boeing Integrated Defense Systems
- General Atomics Aeronautical Company
- General Atomics Photonics Division
- Honeywell Aerospace
- Lockheed Martin Aeronautics Company
- Northrop Grumman Aerospace Systems
- Pratt & Whitney
- Raytheon Company

Technology Horizons

A Vision for Air Force Science and Technology
2010–30

Air University Press Team

Chief Editor
Jeanne K. Shamburger

Copy Editor
Sherry Terrell

Cover Art and Book Design
and Illustrations
Daniel Armstrong

Composition and
Prepress Production
Ann Bailey

Print Preparation and Distribution
Diane Clark

- Rolls Royce Liberty Works
- VMware

Other

- Air Force Red Team
- Air Force Studies Board (AFSB)
- Office of Net Assessment (ONA)
- Office of Naval Research (ONR)
- Naval Research Laboratory (NRL)